Nature's Keeper

In the series

Ethics and Action

edited by Tom Regan

Nature's
Keeper

Peter S. Wenz

Temple University Press

Philadelphia

Temple University Press, Philadelphia 19122
Copyright © 1996 by Temple University
All rights reserved
Published 1996
Printed in the United States of America

♾ The paper used in this publication meets the requirements
of the American National Standard for Information Sciences—
Permanence of Paper for Printed Library Materials,
ANSI Z39.48-1984

Text design by Erin New

Library of Congress Cataloging-in-Publication Data

Wenz, Peter S.
    Nature's keeper / Peter S. Wenz.
        p.   cm.—(Ethics and action)
    Includes bibliographical references and index.
    ISBN 1-56639-427-9 (cloth: alk. paper). —
ISBN 1-56639-428-7 (paper: alk. paper)
    1. Environmental ethics.   2. Human ecology.   I. Title.
GE42.W46   1996
179'.1—dc20   95-47189

This book is dedicated to the five of us

Contents

Acknowledgments

My intellectual debts are too great to remember, much less specify, in detail. I thank the following people who read the manuscript in whole or in part and made helpful suggestions: George Agich, Robert Becker, Harry Berman, Kenneth Berryman, Meredith Cargill, Alex Casella, Ed Cell, Royce Jones, John Otranto, Richard Palmer, Tom Regan, Larry Shiner, Bethany Spielman, Eric Springsted, James Stuart, and Patricia Wenz.

I thank Klaus Biegert and his staff for organizing the World Uranium Hearing, Wendy Moen for inviting me to it, and Drs. Christa and Peter Jecel for hosting Patricia and me during our stay in Salzburg.

I thank the Center for Legal Studies at the University of Illinois at Springfield for supporting my research.

Finally, I thank J. Baird Callicott for his work on Aldo Leopold's land ethic and Karen Warren for the ecofeminist perspective that pervades this work.

Nature's Keeper

The fact that the Jew still lives among us is no proof that he also belongs with us, just as a flea does not become a domestic animal because it lives in the house.
Joseph Goebbels
Nazi Minister of Information

Kills common household pests.
Black Flag Insect Killer

As children, we were taught to respect Mother Earth, all animals, trees, birds, and water. . . . Aboriginal people since the beginning of time never believed they were supreme over all animals. . . .
Lorraine Rekmans
Anishinabe Nation

Establish and extend the power and dominion of the human race itself over the universe. . . . Recover that right over nature which belongs . . . by divine bequest.
Francis Bacon
Father of Modern Science

Currently, reproductive organ cancer among Navajo teenagers in the United States is seventeen times the national average. Worldwide, most uranium mining for nuclear power is done near indigenous people.

So the Indians who live . . . around that [uranium] mine didn't live very long. I lost my wife with cancer, I lost another daughter that had cancer.
George Blondin
Elder of the Dene Nation

In the fission breeder man [has] another path to salvation, . . . an . . . inexhaustible energy source that would, to use H. G. Wells' words of 1914, "Set Man Free."
Alvin M. Weinberg
Nuclear Physicist

In this book I attempt to understand, and suggest how to curtail, the tragedy I was taught to call progress.
Peter S. Wenz

Introduction

Flying on Faith

I always worry that I won't return alive when I travel by plane. At 37,000 feet it was a little late to consider alternatives. Patricia seemed comfortable one row back, except for intermittent shallow coughs.

We were returning from two weeks in Europe, one in Bavaria and the other in Salzburg, where we attended the World Uranium Hearing. At the hearing indigenous people from around the world explained how commercial-industrial cultures ruin native societies. Powerful images of native life being destroyed challenged assumptions of progress that permeated my education. I came to see the Holocaust, the deliberate killing of six million Jews under Nazi control, not as an aberration, but as expressing persistent aspects of our culture's approach to reality. As these ideas flooded in I took notes hoping to make sense of them later. The results of subsequent thought and reading are before you.

I take challenges to my views seriously because I know most beliefs are just articles of faith. How do I know this flight is safe? Hundreds of people and tons of metal move through thin air at 500 miles per hour more than seven miles above solid ground. How is this safe, or even possible?

From what I remember of science the story seems to be this: The air is composed of molecules, which are particles too small for anyone to see, but scientists know that they are there. If the plane moves fast enough through these particles, the curve of the plane's wings causes particles underneath to push up on the wings, resulting in flight.

Of course, I cannot verify that this is how planes lift off and usually arrive without mishap at their appointed destinations. I can say only that this is what people in my society accept as the explanation. And planes do fly, even crowded planes such as the one that carried Patricia and me.

I often wonder how we know what we think we know, especially when what we know seems to violate common sense. Galileo "proved" at the dawn of modern science that the earth travels around the sun, and that the alternations of day and night are caused by rotations of the earth on its axis.

This implies that the earth's surface is moving at almost 1,000 miles per hour where I live, in Springfield, Illinois. Yet I feel no movement and the old medieval story that the sun, moon, and stars do the moving seems obviously true, at least as judged by what we all see everyday. (I've checked it out.)

I do not think Galileo was wrong. After all, people traveled to the moon using these scientific theories. But my acceptance of them is based primarily on faith in authority.

I grew up during the McCarthy era when people were afraid of communism. There was great controversy about whether oaths of political allegiance should be required of teachers, other workers, and students. I realize now that a more fundamental loyalty oath went unchallenged—the confession on science exams of belief in molecules, electrons, and gravity.

We are all in the same position, even scientists. For example, geneticists using an electron microscope or a mainframe computer to map a species' genetic code are unlikely to be experts in subatomic physics, engineering, or computer science. They must rely on experts in these fields to justify using accepted tools of research.

Everyone's acceptance of our society's view of reality is based largely on faith in authority.

I become aware of my faith when doubt creeps in because things do not add up. We were told that school was encouraging us to think for ourselves and that Western science differs from religious, superstitious, and primitive beliefs because it is based on empirical evidence rather than authority. My experience belies both those claims. Our society's self-concept that belief is based on personal experience is simply not true. This puts me on notice that much of what I accept warrants careful scrutiny. Perhaps progress is different from what I was told.

All of my beliefs could not be wrong, however. Planes fly, Americans landed on the moon, and the earth rotates on its axis. At least I think so.

A Call to Hear

The trip to Salzburg was confusing from the start. I spent six days hearing indigenous people talk about their lives. Why me?

I was in my office one morning in the middle of May when, shortly before I was due to start teaching, I received a telephone call from Wendy Moen in Munich, who invited me to the World Uranium Hearing to be held in Salzburg, Austria, from September 13 to 19. She told me the hearing was designed to give indigenous people around the world the opportunity to explain the impact of uranium on their lives and communities. Uranium is mined largely in areas designated for indigenous people, such as on Indian reservations in the United States; nuclear weapons have been tested on or near their land and nuclear waste may be stored there as well. At the hearing indigenous people would speak for themselves and there would be lectures on related legal and scientific topics. Because I had written a book, *Environmental Justice,* I was asked to join the Board of Listeners, whose job was to publicize what we would hear.

People against Nature

Many indigenous people whom I heard speak in Salzburg challenged the belief, common in our culture, that nature (meaning in this context everything nonhuman on earth) exists to serve human beings. This belief comes to Western industrial people (us) from several authoritative sources. The Bible's Genesis story includes the command by God to "be fruitful and multiply, and replenish the earth, and subdue it: and have dominion over the fish of the sea, and over the fowl of the air, and over every living thing that moveth upon the earth." Many people interpret this to require using the earth exclusively for human good.

Ancient Greek thought is also influential in our culture, and the Greeks thought people were of particular value because people alone possess reason. Aristotle, for example, thought that God was pure reason. Because people are the only beings on earth with reason, they are the most godlike, so the rest of nature should be used to serve human beings. The ancient Roman Stoic, Cicero, believed similarly that because animals lack reason, people owe nothing to them.

The German philosopher Immanuel Kant maintained that because animals lack reason, torturing them is wrong only because "he who is cruel to animals becomes hard also in his dealing with men." Practicing kindness toward animals has as its sole object fostering kindness toward people. Animals do not count at all for themselves.

This view is well represented in the twentieth century and is often referred to as "anthropocentrism," the centering of value on human beings. John Passmore, an academic philosopher from Australia, concludes in *Man's Responsibility for Nature* that the only reason to avoid harming nature is to avoid hurting people. An implication is that when people are protected, nature may be harmed. On this view, saving the spotted owl *for itself* is no reason to restrict logging old growth forests in the U.S. Northwest.

William F. Baxter is an economist who favors free market approaches to environmental issues. In *People or Penguins: The Case for Optimal Pollution*, he writes, "every person should be free to do whatever he wishes in contexts where his actions do not interfere with the interests of other human beings." So if DDT damages penguins, but not people, there is no problem. "Penguins are important because people enjoy seeing them walk about on the rocks." They are not important for themselves. Furthermore, Baxter contends, "No other position corresponds to the way most people really think and act—i.e., corresponds to reality."

This view is debatable. Near the beginning of our century Henry S. Salt, a pioneer of ethical vegetarianism, questioned the biblically inspired view that animals were given to people for human use. He claimed that he had looked, but was unable to find the receipt. In our own day Australian philosopher Peter Singer questions anthropocentrism in the name of animal welfare, endorsing what he calls "animal liberation."

Others take a broader approach. They reject anthropocentrism in the name of nature in general, not just animals. This was the view of the many indigenous speakers in Salzburg who advocated respecting all of nature, and offered prayers to Mother Earth. Their view resembles Aldo Leopold's land ethic and the deep ecology movement, which accord moral importance to all of nature, not only to individual human beings or other individual animals.

I find this attractive: People should respect nature for itself, not just use it for the good of human beings. But how can I convince others? Maybe some people see intrinsic value in nature and others do not, the way some people see a point to golf and others do not, or the way some people like chocolate and others do not. This would explain the apparent stalemate in modern culture on this matter. Individuals with opposite views seem unable to convince one another and philosophers are enjoying little success distinguishing this issue from one of individual preference or taste. How can progress be made?

An Indigenous Perspective

Help came during the second day of the hearing from an indigenous person, Lorraine Rekmans of the Anishinabi Nation of northern Ontario, who told a story to illustrate the perspective she wanted us to share. Her six-year-old son had a science project for school, growing grass in a small container to learn what factors affect its development. The project seemed to be going well. But one "morning he had a very concerned look on his face, and he said, 'Mom, I have to go water my grass because I can hear it screaming.'"

Screaming grass! Was this childish fantasy or profound insight? Perhaps people who can hear the grass scream are reluctant to mow it down, tear it up, and genetically alter its seed. On the other hand, those who consider grass to be at our disposal may be led by degrees to similarly view insects, mice, cattle, and deformed human beings, connecting our culture's anthropocentrism to the Holocaust.

African, Canadian, Indian, Ukrainian, and many other witnesses pointed out at the hearing that people who are not oriented, as indigenous people are, toward reverence for the earth, people who cannot hear the grass scream, are likely to miss human screaming as well. Those who think it is their right or duty to subdue the earth tend to subdue other people in the bargain. This was the perspective that promised a philosophical breakthrough.

I think indigenous people at the hearing were directing their comments at European-inspired commerce and industry—*our* culture—because it is our culture that spawned the use of nuclear and other technologies that disrupt their lives. I think they were saying that our culture subdues people along with nature.

This is strange, because our culture often justifies manipulating nature as a means of advancing human well-being. Nevertheless, I came to realize our culture inevitably oppresses people in the course of manipulating nature for the good of humanity.

This led me to question alternatives. Are there societies that oppress people less than we do? Indigenous people at the hearing suggested that many tribal societies that use simpler, less powerful technologies are less oppressive than ours. Although some are violent, such as the Yanomamö of northern Brazil, who regularly ambushed and killed men from neighboring villages, other tribal people, especially foragers (hunter-gatherers), oppress people much less than we do.

Is it possible to live an attractive, truly human life in such societies? I was taught that people naturally want as much material comfort as possible and manipulate earth maximally to achieve it. Indigenous cultures with simple technologies that affect nature minimally seem pitifully poor and unattractive. However, Lorraine Rekmans and other indigenous speakers at the hearing maintained that many native people enjoy riches that industrial people cannot understand. Indigenous life in societies with little human oppression is desirable.

I wondered how nature is viewed in these societies. Do people there reject anthropocentrism and does this rejection support low levels of human oppression and high levels of cultural riches? Yes. People in these cultures have a social conception of relationships between people and other natural constituents, both animate and inanimate. They include all constituents in their concept of community and view consideration, respect, and reciprocity to be appropriate in all dealings with fellow community members. This view pervades these cultures, affecting the treatment of both people and nature.

According to author Jamake Highwater, "When the Indian potter collects clay, she asks the consent of the river-bed and sings its praises for having made something as beautiful as clay. When she fires her pottery, to this day, she still offers songs to the fire so it will not discolor or burst her wares."

Ms. Rekmans, after telling the story of her son hearing the grass scream, introduced her Ojibwa name, Peh-sheh, which means robin. "I am all things that the robin is," she said.

> This is significant because it recognizes that animals have characteristics, and this is how I find my place in the universe by identifying

with one other being of nature. This is how I always remember that I am not superior to the other beings and living things on this planet.

Indigenous people typically consider themselves "only part of the intrinsic circle of life."

Another native speaker, Esther Yazzie, concurred. She attended the hearing with her husband Robert and teenage daughter Darnell in order to share a Navajo perspective. Coming from the four corners area of the United States, she found it easy to relate to the Austrian Alps. "Mountains are made by Mother Earth," she said. "Mountains were brought up by the center of Mother Earth. They are the grandparents, and we are the grandchild." She concluded with a poem expressing a social view of nature:

The new way to exist on Earth
may well be an ancient way
of steadfast lovers of this particular land.
No one has better appreciated Earth
than Native Americans.
Whereas to the White Man,
only the white attains full human status,
everything to Indians are a relative,
everything was a human being.

Can we profit from knowledge of such indigenous cultures? Given our historical, cultural, economic, and technological contexts, we certainly cannot replicate such cultures and become indigenous ourselves. We must face the future from where we are, sitting behind computers eating fast-food lunches. Even if some indigenous cultures with a social conception of nature fulfill human aspirations and treat people better than we do, it is unclear whether hearing the grass scream is even possible, or helpful, for us.

The Paradoxical Thesis

Reading and reflecting since the hearing has led me to the paradoxical conclusion that, in our cultural context, *attempts to master nature in the human interest result in human oppression.* This is my paradoxical thesis. Its corollary is this: *A necessary component in changes that reduce the human suffering caused by our society is adopting the perspective of some indigenous cultures insofar as they value nature for itself.* When we try to serve only people, our historically evolved ideas, practices, and institutions influence us to oppress many among us. Paradoxically, the harms that our culture inflicts on people result largely from our culture's exclusive orientation toward human good.

Suppose for a moment that this is true. Then the debate between those who believe the earth exists exclusively for human good and those who respect the earth for itself can be concluded. If the people who value only human beings were convinced that the only way, in our cultural context at least, to serve people well is to respect nature for itself, they would endorse such respect. In other words, if the paradoxical thesis is correct, the terms of debate would have to be altered because opposition between people and nature would misrepresent the options available. The option of respecting both people and nature would exist, and be superior.

The situation would resemble the opposition between leisure and work for workaholics. Whereas many people work hard to be able to afford leisure activities, some people enjoy their work so much that leisure activities are not desired. For these people it would make no sense to speed up, move, or in any other way alter their work to have more time to enjoy their leisure, because their work activities are the most enjoyable they know. Isaac Asimov, for example, an extremely prolific writer, quipped that he wrote so much because he was too lazy to get up from the typewriter. For people like this, the opposition between leisure and work would not be helpful. Similarly, the opposition between people and nature would not be helpful if it turned out that, paradoxically, the best way to serve people is to avoid trying to serve *only* people. The debate between "people first" and "nature above all" advocates would be concluded not by one side convincing the other, but by depriving the disagreement of any point.

A comparison with people's love of their children may be helpful. I remember seeing an ad when I was a child for pairs of rabbits for breeding. It recommended raising rabbits "for fun and profit." There was a clear distinction between people and rabbits. The rabbits were the means toward the goal of improving the lives of people. But we do not typically think of children that way. We ideally think of children as ends in themselves. We have children, raise them, educate them, and are kind to them because we love them *for themselves*. Of course, interacting with children in these ways can be, and often is, fulfilling, meaningful, and fun. At other times it is frustrating, annoying, and inconvenient. The point is that even when childrearing is fun, the fun depends on the relationship being guided *not* by the prospect of fun or fulfillment, but by love of the children for themselves.

There are people whose choice to have children is guided by their own needs for fulfillment. Those who remain fixated on their own needs, whether for fulfillment, variety, or fun, are likely to be disappointed. This is the paradox: It is self-defeating to think of children as the means toward the end of parental fulfillment, even though raising children is often rewarding for parents. Similarly, it is self-defeating to think of nature as a means to achieve human welfare, even though nature does serve human welfare.

A spiritual dimension may be important as well. Indigenous people at the hearing had a sacred (as well as a social) view of nonhuman nature. Ms. Yazzie said that the Navajo people consider their mountains to be sacred and because they are sacred, "The Navajo people place their offerings on . . . mountain tops" and address the mountains with prayers, songs, and ceremony. Ms. Rekmans spoke in a similar vein:

> I would like to explain to you the significance of the sacred circle as told to me by my grandfather. As children, we were taught to respect Mother Earth, all animals, trees, birds, and water. We were taught to take from Mother Earth with reverence and offer gifts and thanks to show our appreciation for the life-giving things earth had provided.

People in our society could benefit from the spiritual attitude of those people who revere the earth, in whole and in its various parts, as sacred. People with this religious attitude do not try to subdue nature. If the paradoxical thesis is correct, the result in our cultural context would be the better treatment of people. This is a secular reason (based on the value of respecting people) for a religious attitude toward nature, an attitude that is compatible with any of the world's great religions, including Christianity.

The paradoxical thesis supports a position quite different from Immanuel Kant's. Kant opposed cruelty to animals to avoid cruelty to people, but he rejected valuing nonhuman animals for themselves, as we regard human beings. By contrast, I advocate valuing nonhuman animals and other parts of nature for themselves. So long as we (in our cultural context) view nature the way Kant viewed animals, i.e., merely as a *means* to promote human well-being, human oppression will result. Consequently, nature should be valued for itself to the point of developing a spiritual or religious attitude of reverence.

The Intellectual Journey

Why should anyone believe anything so paradoxical as the claim that our culture's dedication to the service of people results in human misery, whereas some cultures that value nature for itself, and revere it spiritually, are more likely to treat people better?

The answer lies in history. I do not know if I would have realized this if the hearing had not been in Europe, where people are surrounded by evidence of the history that has shaped Western industrial cultures. But in that setting it came to me that if the paradoxical claim is true, it should be evident in the continuing influence of historically important ideas, movements, and practices that were infused with the idea that only human beings are ends in themselves. Four of these strike me as most significant. They are, roughly in the order of their historical appearance: Christianity (just one, particularly influential, version), commercialism, industrialism, and modern bureaucracy.

I describe individually each idea, movement, or practice, but emphasize their joint and continuing influence on our society. If each description can be compared to a picture, then I intend the picture to be like a series of transparency overlays in a medical text. Each is made transparent because a better picture of the entire organism is given when the individual pictures can be viewed simultaneously as well as individually. Similarly, although each idea, movement, or practice will be discussed individually, I hope that a concept of contemporary Western culture will emerge as the simultaneous consideration of these practices yields composite understanding. Altering the analogy from medical texts to textiles, each of the four types of influence is treated here as an individual strand in the fabric of contemporary society. I want to expose the pattern that results from the strands' interweaving *in contemporary society*.

The general pattern includes five interrelated elements: 1) people conceive of themselves as essentially separate from nature; 2) people think they are in some kind of jeopardy; 3) subduing nature is undertaken to reduce the (perceived) danger, but usually augments human jeopardy; 4) because people are increasingly in (perceived and often real) jeopardy, they form increasingly centralized and powerful concentrations of force; 5) they use this force *against people* believed either to precipitate danger or to inhibit its reduction. I call this the five-part pattern.

The historical narrative is designed to highlight twentieth-century manifestations of this pattern. I pay special attention to the plight of indigenous people and to the Holocaust, which I find related both to one another and to enduring features of Western civilization.

The second half of this book includes information gathered at the World Uranium Hearing about the cruel irrationality of our mining, use, and disposal of uranium. Because the inhumanity of killing people with nuclear weapons is obvious, I discuss primarily the use of fission to produce electricity. The development and use of nuclear power illustrates the same five-part pattern. Begun as an attempt to gain more for people from nature, the production of nuclear power results in human oppression. The words of victims who testified at the World Uranium Hearing are grave and powerful. My account ends with reflections on the Gulf War of 1991. I then discuss indigenous cultures in which people are less oppressed than in our own and life seems culturally rich and generally fulfilling. I show that integral to these cultures is a social and reverential attitude toward nature that rejects increased mastery in the human interest.

The final concern is whether consideration of these indigenous societies can help us to improve our own. Frankly, I do not know. I am sure that we cannot fruitfully imitate, and so become, indigenous people. If we incorporate into our culture some insights gained from the study and appreciation of indigenous cultures, including a social conception of and a spiritual attitude toward nonhuman nature, we will create something new, rather than return to some mythical golden age.

Improving our society requires rejecting some widely held assumptions of our age: That a growing economy helps people, that increasing world trade improves standards of living, that genetic technologies are worthwhile, and that family values can thrive alongside commercial growth.

Good reasons are needed to justify working toward a society that rejects many of our cherished beliefs. These reasons must be based on other beliefs that we hold. Conclusions drawn from radiometric data, for example, will not convince a creationist that the earth is many millions of years old, unless the creationist accepts a host of beliefs about radioactive particles, half-lives, measurement techniques, and so forth. Similarly, reasons to reject anthropocentrism in our culture must employ other beliefs *in our culture.* The present work features beliefs about history, the current state of human welfare, the desirability of reducing human oppression, and the importance of promoting family values and genuine human fulfillment. Unless we start valuing nature for itself, human oppression will increase. But if we adopt this value, we *might* realize improvements in personal security, interpersonal relationships, and human fulfillment.

Chapter I

Our Christian Heritage

Plague and Passion Play

Patricia and I began our European adventure with a week's vacation in the Bavarian Alps before the hearing started. I never guessed this first week would profoundly affect my understanding of hearing lectures and testimony.

Our first stop in the Alps was Oberammergau, the town famous for its passion play that is performed there all summer every ten years, in the first year of every decade. The play depicts the last days of the life of Jesus, from the Last Supper through the Crucifixion to the Resurrection. Because it was 1992 the play was not being performed while we were there, but the town's renown for its passion play reaffirmed for me the importance of Christianity in European history and reminded me of a famous article by Lynn White Jr.

White advanced the controversial thesis that Christianity caused the environmental crises of Western civilization. Not being a theologian, or even a Christian, I cannot evaluate this charge. However, I can say that some aspects of Christianity have been appropriated in recent centuries by antienvironmental thinkers.

As often interpreted, Christianity supports the first two elements of the five-part pattern—the separation of people from nature and the existence of human jeopardy. Christian history in the Middle Ages illustrates additionally elements four and five—centralization of power in the face of jeopardy and the use of that power to oppress people considered dangerous. Christianity is not essentially oppressive or antienvironmental. But some people have justified oppression and antienvironmentalism in its name, and it is useful to see the aspects of Christianity used (mostly misused) by these thinkers.

Anthropocentrism and Original Sin

The Old Testament is sometimes interpreted to support anthropocentrism. God said after creation that the world is good. But the command to subdue the earth and have dominion over every living thing often has been taken to mean that God created the world for no reason other than to serve human beings. The world is good in order to support humanity.

Christians also often assume that only human beings have an afterlife, with rewards in heaven and punishments in hell. If this life is but a preparation for the next, which is more significant, then perhaps God created the earth and all its nonhuman creatures only to serve as a testing ground for people. The earth and its nonhuman creatures are not important for themselves.

But why are people so important? People alone have free will, the ability to choose right from wrong. This is why it makes sense to put people to tests of good and evil, whose results determine rewards and punishments. People deserve the consequences of their behavior because, given free will, they are responsible for their actions.

This account of good and bad behavior is complicated, however, by the doctrine of original sin. According to the Old Testament story, because Adam and Eve disobeyed God by eating the fruit of the forbidden tree, people's souls are inherently corrupt, inclining them toward evil instead of good. This puts them in cosmic jeopardy of eternal damnation.

Many of people's tendencies to evil concern relationships to material beings. For example, people often seek material riches instead of righteousness, and temptations of the flesh can lead to sins of gluttony and adultery. In order to avoid eternal punishment, then, people must resist these earthly temptations. So the material earth, although good when God created it, is now a snare for people, a temptress that people must guard against.

There is still nothing antienvironmental here. Although the earth is relatively unimportant, on this view, there is no justification for its degradation. In fact, most of the incentives in our culture for mastering, controlling, and altering the earth stem from desires to make life on earth longer, more comfortable, less painful, or more fun. These incentives are excluded from Christian thinking that emphasizes rewards in heaven and punishments in hell.

Nevertheless, according to Father Thomas Berry (a Christian environmentalist), an overemphasis on redemption and the promise of the hereafter detracted from Christian celebration of the earth and all its constituents as God's wonderful creation. "The natural world is the larger sacred community to which we belong. . . . If this sense of the sacred character of the natural world as our primary revelation of the divine is our first need, our second need is to diminish our emphasis on redemption experience in favor of a greater emphasis on creation processes." However, he notes, "Excessive emphasis on redemption, to the neglect of the revelatory import of the natural world, had from the beginning been one of the possibilities in Christian development. The creed itself is overbalanced in favor of redemption." This overemphasis on redemption

> continued through the religious upheavals of the sixteenth century and on through the Puritanism and Jansenism of the seventeenth century. This attitude was further strengthened by the shock of the Enlightenment and Revolution periods of the eighteenth and nineteenth centuries.

Thus, the version of Christianity emphasizing cosmic insecurity and the need for redemption and salvation became influential early and continued to dominate. While it was not essentially antienvironmental (it did not support despoiling the earth), it was anthropocentric and did not encourage celebrating God's creation.

Insecurity Breeds Concentration of Power

Medieval Christianity illustrates how insecurity among the masses can lead to an increased concentration of power and the oppression of human beings. This is not to say that Christianity, when properly interpreted, calls for witch burning, inquisitorial torture, or any number of atrocities that have been committed in its name. I cannot imagine that the gentle man whose life story is told in the Gospels would torture anyone. The fact that brutalities were nevertheless committed in his name suggests a strong tendency for insecurity to produce power concentrations and human oppression.

Security considerations often justify the concentration of power in relatively few hands. George Orwell indicated this in *1984* by having The Party maintain constant war. Because the country was at war, measures repressing dissent were justified.

In medieval times, when people felt insecure about the effects of original sin on their afterlife, they relied on Church sacraments, such as baptism, communion, and last rites, to abolish sins and foster heavenly reward. So Church officials in control of the sacraments had great power. Insecurity in the face of jeopardy justified the concentration of power.

Of course, cosmic jeopardy was not the only insecurity during medieval times. There were diseases, bandits, and wars, as there are today. Diseases were handled by priests as well as by doctors, because doctors were often ineffective. A notorious example was the Black Death that afflicted Europe intermittently, starting in 1348, for nearly 300 years. Perhaps one of every three Europeans died of this plague in just the first three years. And the worst was yet to come. Historian Barbara Tuchman notes that the disease was widely assumed to be "a chastisement from heaven."

The passion play for which the town of Oberammergau is justly famous reflects a theological interpretation of the disease. It began as an offering of thanksgiving to God for the fact that in 1633 the Black Death stopped just short of the town. People's insecurity in the face of some dangers of this life, as well as their insecurity in the face of the afterlife, resulted in the Church acquiring concentrated power, as people looked to the Church for protection.

Counts, dukes, princes, kings, and other members of the aristocracy were supposed to reduce the dangers of theft and war. Here, again, the existence of insecurity justified concentrating power in relatively few hands. Despite clashes between secular rulers and Church officials, cooperation was the norm.

Once instituted, power concentrations of these kinds tend to continue, as most people with this power want to keep it. People in medieval times found that emphasizing insecurity was an effective way of attaining and keeping power.

The same is true in our own time. In the United States, the perceived threat of international communism was at the heart of the cold war that lasted more than forty years. Due to this (perceived) threat, Senator Joe McCarthy was able in the early 1950s to wield enormous power, and abrogate the freedom of many Americans, who were denied gainful employment on grounds of political affiliation.

There are as many justifications for concentrating power and surrendering freedom as there are insecurities. People are worried today about the widespread use of illegal drugs, so they accept questionable police practices. Radio reports of drug interdiction in Illinois suggest that cars with Texas and Florida license plates driven by people with Hispanic surnames are often stopped and searched because the drivers allegedly made improper lane changes. Because significant quantities of illegal drugs may be interdicted in this way and drugs cause insecurity in our society, people acquiesce in these questionable searches.

Similarly, President Clinton's reaction to the bombing of a federal building in Oklahoma City in April 1995 was to propose new powers facilitating FBI infiltration of dissident groups. Such powers had been abused during the civil rights and Viet Nam War eras, when the FBI violated the privacy of Martin Luther King and some U.S. senators on bogus grounds of U.S. security. The relationship of insecurity to concentrating power and relinquishing freedom extends from political and legal contexts to personal ones. Most people in our society hope to live long, healthy lives. Ill health jeopardizes what we value highly, so people in our culture have traditionally invested great faith and power in physicians. Physicians are among the highest paid service workers. They largely regulate their own profession. Most significant, their recommendations are still often called "doctor's orders," and it seems the sicker people are, the more likely they are to view doctors as authority figures. Finally, when they do question a physician's judgment they typically consult another physician. The insecurity of ill health motivates people to put power in the hands of doctors.

Medieval Repression of People

The cosmic insecurity entailed in the doctrine of original sin motivated human oppression in the Middle Ages. To receive eternal life, people must relate to God properly, and the Church is needed to mediate the relationship. In addition, people must believe and act according to the better part of themselves (spirit rather than matter), orienting their lives toward heaven and eternal afterlife. This is not easy because original sin inclines people to the opposite orientation, that is, toward temporal, earthly, and material gratification and rewards.

Given this world view, humanity is jeopardized by any groups of people or social forces that weaken the authority of the Church (the needed mediator between God and humanity), or strengthen sinful tendencies (toward material gratification). Such people or forces should be suppressed for the good of humanity.

Consider, first, relationships between men and women. The body's urges tempt people to act sinfully (jeopardizing immortal souls). In a society already dominated by men, women were associated with temptation and the earth. They were considered closer to the earth than men due to their greater material involvement in the processes of gestation and birth. Just as earth is by nature subject to the rule of heaven, women are, according to St. Thomas Aquinas, by nature subject to the rule of men.

As for temptation, it was Eve who prompted Adam to disobey God. Tuchman writes, "Woman was the Church's rival, the temptress, the distraction, the obstacle to holiness, the Devil's decoy." She was described in the thirteenth century as "the confusion of man, an insatiable beast, [and] a hindrance to devotion." In the fourteenth century a preacher in England warned his congregation:

> In the woman wantonly adorned to capture souls, the garland upon her head is as a single coal or firebrand of Hell to kindle men with that fire. . . .In a single day, she inflames with the fire of her lust perhaps twenty of those who behold her, damning the souls God has created and redeemed at such cost.

When women are this dangerous, their control by men is justified by simple prudence.

This is not to say that suppression of women was due solely to insecurities stemming from fear of sin, but that these insecurities helped to legitimate women's oppression.

Heretics and Jews were suppressed as well. Jews were despised as those who had killed Christ. Pope Innocent III announced in 1205, for example, that Jews were condemned forever to a life of servitude. Jews were despised also because they were a constant reminder that some people could experience Jesus and reject Him. Sociologist Zygmunt Bauman writes:

> The Jews . . . were venerable fathers of Christendom and its hateful, execrable detractors. Their rejection of Christian teachings could not be dismissed as a manifestation of pagan ignorance without serious harm to the truth of Christianity. . . . Jews were . . . people who in full consciousness refused to accept the truth when given the chance to admit it. Their presence constituted a permanent challenge to the certainty of Christian evidence.

Such bad examples had to be suppressed before the temptation to imitate them led to the damnation of people who would have otherwise enjoyed Christian salvation.

The danger posed by Jews is structurally similar to that posed by women. Both constitute a temptation to behavior (indulgence in bodily pleasures, rejection of Christ) that leads easily to damnation. The same applied to heretics, "Christians" who rejected key Church doctrines. The fourteenth century cleric John Wyclif, for example, denied that Church sacraments were necessary for salvation. Widespread adoption of this view would have endangered the Church, its mission of mediation between God and people, and the souls of all who rejected Church sacraments.

The Inquisition was the answer for all such obstinate dissent. Tuchman maintains, "Denunciations, trials, and burnings increased, and in its tortures of supposed heretics the Inquisition was as savage and ingenious in cruelty as any infidel."

Consider next the Crusades, wars on Moslems to regain the Holy Land. "The infidel, like the heretic, was . . . feared as a genuine menace," according to Tuchman. Moslems in control of the Middle East interfered with holy pilgrimages to Jerusalem undertaken by penitent believers. Historian Frederic Duncalf writes, "To go to pray at the Holy Sepulcher was the best of all Christian pilgrimages. The crusaders were fighting pilgrims who set out to open up the route to Jerusalem, which had been obstructed by the Selchukids, and to liberate the holy city." This would improve the access that Christians needed for pilgrimage, prayer, and salvation.

Heathens, people not yet aware of the good news about Jesus or of the bad news about original sin and eternal damnation, were oppressed by Christians trying to save souls. Through Christian domination of their territory, heathens were oppressed for their own good. In 1547 Juan Gines de Sepulveda argued:

> What is more appropriate and beneficial for these barbarians than to become subject to the rule of those whose wisdom, virtue, and religion have converted them from barbarians into civilized men, . . . from being impious servants of the Devil to becoming believers in the

true God? . . . For them it ought to be . . . advantageous . . . since virtue, humanity, and the true religion are more valuable than gold or silver.

I do not suggest for a moment that this is a theologically correct version of Christianity. But it was influential in forming Western industrial culture, and its influence continues. In my lifetime many conscientious Americans wanted to "kill a commie for Christ," communists having replaced, or joined, Moslems in the role of infidel. They were feared on religious grounds (hence, the killing was "for Christ") because they were materialists ("Godless communists"). They denied the spiritual nature of human beings, the existence of heaven, and the reality of God and afterlife.

The fact that the Christian message of peace could be ignored by so many people who were probably attempting to live Christian lives testifies not to the nature of Christianity, but to the influence on people of insecurity. The relationships displayed here between insecurity and both power concentration and human oppression are important because they recur in secular forms in the modern world.

Secularizing Jeopardy and Power

Early modern science played an important part in secularizing Western culture. Seventeenth century scientists, such as Galileo and Newton, explained an increasing range of phenomena by assuming that material reality is composed of atoms that interact mechanically, as in mechanical clocks, and that all mechanical operations can be described mathematically.

We are thoroughly used to thinking of reality in these terms. We believe the material elements are just so many different types of atoms, differentiated from one another by their atomic weights. All other material substances are mathematical combinations of these elementary atoms. In biology, too, mathematics rules. Healthy blood is blood with certain mathematically describable combinations of white blood cells, red blood cells, sugar, and so forth. The heart is like a mechanical pump and health depends on the pressure being appropriate. Some people think the brain is like a mathematically based computer.

If the entire material universe operates by mechanical laws, God's continuing presence is unnecessary. Just as a good watch can run without the watchmaker, the universe can run without God. Scientific reasons for believing in a living God are reduced and the secularization of society is advanced.

Thomas Hobbes reflected the secular turn by applying scientific theories of his day to human society and morality. Hobbes assumed that people are, like all other material beings on earth, just clusters of atoms. Like mindless bowling balls, we tend naturally to continue in a given direction until deflected by obstacles. This means we are inclined by nature to remove all obstacles in our path, and will kill anyone who stands in our way.

According to Hobbes, then, there is no natural milk of human kindness. People are by nature like sociopaths, completely selfish and ruthless in pursuit of their objectives. Because each is dangerous to others, human relationships are like those in a war. There are no moral duties or obligations to other people in this situation, which Hobbes calls the state of nature.

People can escape this dangerous and undesirable situation only by each agreeing to obey the dictates of a single peacekeeper, on the condition that everyone else does the same. So the natural desire for survival, combined with reason, inclines people to establish a *social contract* that requires them to obey on pain of death the dictates of the Leviathan, a single center of power.

There is no right of rebellion against the Leviathan because rebellion threatens to return people to the worst condition of all, the chaotic, anarchic state of nature. With the advent of the Leviathan, then, duty and obligation enter the human social world, as people have a duty to obey the Leviathan. This duty exists merely because the Leviathan has the power to kill transgressors, so, in this world might makes right.

Hobbes's thinking is in some respects a secular version of Christianity. Like original sin, people's sociopathic nature creates jeopardy that warrants concentrating power. Hobbes here gives secular support to the second and fourth elements in the five-part pattern—people are in jeopardy, and concentrating power is a reasonable response to that jeopardy.

Hobbes justified the fifth element as well—the oppressive use of concentrated power. He gave secular warrant for colonial conquests that had been promoted originally as bringing religious salvation to savages. Hobbes's reasoning is that might makes right in this world because morality extends only as far as the Leviathan's power to punish transgressors. Unless and until there is a super-Leviathan whose force extends beyond national boundaries, morality does not govern international relations, and there is no moral reason to refrain from colonial conquest.

Echos of this view persist in the so-called Realist School of foreign affairs, associated with the late Hans Morganthau, in which Henry Kissinger, among others, was educated.

Hobbes's views were controversial due partly to his unconventional value conclusions. His materialist account of human beings also struck many people as implausible. How can material atoms produce thinking? This remains an unsolved mystery.

The Separation of Mind from Body

Early modern thinkers lent secular, scientific support also to the separation of people from nature. Rene Descartes, a younger contemporary of Hobbes, was a mathematician, a scientist, and a philosopher. He agreed with Hobbes that the physical world is like a giant machine whose movements scientists could, in principle, predict, but he maintained that human beings are more than material entities. People can think, and Descartes found it impossible to imagine how purely material entities could do this. He concluded that the human being, the one who can think, must be not the body, but a spiritual mind or soul that thinks, wills, considers, feels, dreams, believes, and knows. The material body is just a machine that the person (the mind) uses during this life.

According to Descartes, human beings are the only earthly creatures with minds. Everything else, from rocks to dogs, is just mindless matter. Only people, then, are subjects; everything else on earth is just an object. Descartes here gives secular support to what was originally a religious view (the first element in the five-part pattern) that human beings are essentially separate from nature. This helps the pattern persist in secular environments.

Because only people have minds, Descartes reasoned further, only people can experience pleasure and pain. Nonhuman animals can feel no pain because they are machines without minds. This is why it is permissible to practice vivisection, cutting into live animals for scientific experiments. The animals feel no more pain than would any other machine, their "screams" being like the sound chalk makes when scraped against a blackboard. In this way, Descartes supports the third element in the five-part pattern, subduing nature in the (supposed) human interest.

Immanuel Kant, writing 150 years after Descartes, also gave secular reasons for distinguishing people from nature. Everything in the mechanical universe of material things operates by invariable laws of nature. In principle, then, all material events can be predicted in advance. But people have free will and are for this reason unlike any other beings on earth. People also have reason, which enables them to understand their moral duties. On these grounds, Kant concluded that humans are the only beings in the world that have value for themselves. Everything else is a means to the end of human fulfillment.

Kant disagreed with Descartes, however, about the ability of nonhuman animals to experience pain. Somehow, he thought, animals can experience pain even though they are material beings lacking free will and reason, and so cruelty to animals is therefore possible. It should be avoided, Kant thought, because it leads to cruelty toward people, but he concluded that using animals for human welfare is not cruel.

Millions of animals currently are used each year to help people. Rabbits are deliberately blinded by having chemical solutions dripped onto their eyes to assure the safety of new cosmetics. Millions of furry animals are caught in painful leg-hold traps to make garments. Countless hens are confined for their entire adult lives in small wire cages to reduce the price of eggs. None of this is immoral on the assumptions that people are essentially separate from nature and that people alone are morally important, views given secular support by Descartes and Kant.

Chapter 2

Commercialism

The Five-Part Pattern in Commercialism

After visiting Oberammergau Patricia and I went to Mittenwald, which *Fodor's* describes as possibly "the most beautiful town in the Bavarian Alps." It is on a north-south trading route dating from Roman times, but its wealth dates from the late Middle Ages when it became a link in the commercial connection of Verona to Munich.

Trade is an ancient activity. Before the Romans, trade was carried on by Egyptians, Chinese, and Persians, to name but a few. The Middle Ages, however, were a time of reduced trade in Europe. Current patterns of commercial activity can be traced to the resurgence of trade after the Crusades.

The story of commercialism's growth and current conduct helped me to formulate the paradoxical thesis, the view that our culture's attempts to serve exclusively the human good by dominating nature result in human oppression. Commercialism illustrates the five-part pattern: People think they are essentially separate from nature, and in jeopardy. They subdue nature to reduce perceived dangers, but instead augment human jeopardy. Responding to increased jeopardy, people create more power concentrations, which they use increasingly against fellow human beings.

This pattern can be seen in history and in current world conditions. The following information is a sample that highlights the oppression of indigenous people.

First Part: Separation from Nature

The self-perception of being separate from nature was inherited from both the Judeo-Christian religious tradition and ancient Greek philosophy. Later thinkers, such as Descartes and Kant, reinforced this perception with modern, secular, philosophical justifications that stressed the difference between mind and body in the individual human being. They associated the mind with the distinctively human and the body with nature, since many animals have bodies much like our own. Just as the mind should rule the body, humanity should rule nature.

Second Part: Human Jeopardy

Human jeopardy results from scarcity, according to working economist Clement A. Tisdell: "The basic [economic] problem is considered to be how to manage or administer resources so as to minimize scarcity, that is, the 'gap' between individuals' demand for commodities and the available supply of them."

This scarcity, like original sin in Christianity, is a condition of jeopardy that should be overcome for the good of human beings. Extracting as much as possible from the earth for human use is the best way to combat scarcity. "The scarcity problem explains the interest of economists in the efficiency of alternative social mechanisms for resource allocation and their interest in economic growth and in development." People who really love humanity will not restrict needed manipulations of nature.

Secular bases of value support the view that scarcity is endemic to the human condition. When the scientific view of the universe as an enormous mechanism weakened justifications for belief in a living God, it also weakened belief in God-given values. Philosophers were called upon to explain what is good and valuable without divine commands.

The task was complicated by another implication of science—the absence of objective values in nature. Science investigates objective reality and finds no values there. For example, we can describe the destruction of the spotted owl's habitat by continued logging of old growth forests, but that description does not by itself tell us whether to stop the logging in order to save the owl. David Hume is credited with popularizing the claim that no investigation into nature or physical reality could ever result in conclusions of value or in prescriptions (statements of what people ought to do).

Philosophers have not yet solved to their general satisfaction the problem of locating a source of values other than God and nature, but one recurrent tendency is to identify the source with human beings. Philosophers tend to associate the distinction between values and facts with that between subjects and objects. Science investigates objects and finds neither values nor prescriptions. Values and prescriptions come from people acting as subjects. Values come into existence when people value things and prescriptions come into existence when people prescribe things. Issuing exclusively from people, values and prescriptions are in some sense subjective (the exact sense being highly controversial), whereas facts and descriptions are objective.

If values are the creation of human subjects, who can rely on neither science nor religion, how can people decide what to value? Jeremy Bentham maintained that people have only one ultimate value—pleasure. People who value snow for skiing do so because skiing gives them pleasure. Those who value large meals do so because they gain pleasure from eating. Others who diet derive pleasure when they look in the mirror, or when others look at them with admiration. Pleasure is behind every value, Bentham claims. Even when we experience value conflicts, for example, between dieting and eating dessert, the conflict is about which course will yield the most pleasure.

This view reinforces belief in the existence of scarcity. According to Bentham, whatever gives pleasure is good and worthy of pursuit, insofar as it gives pleasure. Because all desires have this same ultimate basis in pleasure, no distinction exists between wants and needs. People are free to consider all their desires (for pleasure) to be needs worthy of fulfillment. Because people's desires are unlimited, needs become unlimited, and the problem of scarcity (unfulfilled needs) is aggravated. If scarcity is a condition of human jeopardy, then jeopardy is permanent.

Third Part: Subdue Nature to Reduce Jeopardy

Specialization helps to maximize nature's yield. When people specialize they can become more expert and efficient at manipulating nature to serve human beings. People who specialize in home construction, for example, are more likely to make better homes and use available materials more efficiently than others can. The same is true of most production.

When people specialize in production, they must trade with one another to obtain most of what they need. Home builders trade with farmers for food and with tailors for clothing. This is commerce, which is a necessary means of reducing scarcity, because it is a corollary of specialization.

Specialization and trade require increasing production for exchange, rather than for use. Before the modern era and the rise of commercialism in Europe, most Europeans lived in a subsistence economy, as do many indigenous people today, where almost everything people use or need is produced locally using local materials, and money plays little part in daily life.

Most subsistence economies are also commons-oriented. People share much of what is needed for material well-being without anyone owning it individually. In medieval England, for example, much of the land was used as a commons, although part also was owned individually as garden plots. Everyone had access to the commons to graze animals and many people grew vegetables there instead of in private gardens. Peasants had the right to gather fuel wood and peat from the commons. They also could put out cattle on the pastures and pigs in the (common) forests.

Values in such societies depend on use. Simply living requires the use of many things, such as food, water, shelter, and clothing, which are valued simply for the uses people make of them. But when people specialize in production, they must exchange what they produce for most of what they need. They view what they produce not so much for its value in use as for its value in exchange for other things. We are so accustomed to an exchange economy that when asked how much something is worth we usually think in monetary terms, which is value in exchange. More than two hundred years ago Adam Smith *defined* economics in terms of exchange values.

Northern Italian states, especially Venice, were early centers of the trade that resulted in greater emphasis on exchange values. Receiving goods by sea from the Near East, they shipped them by land and river to other parts of Europe. Mittenwald prospered in the era of early medieval European commercialism because it was situated on a major trading route. Goods shipped overland from Verona could be packed onto riverboats at Mittenwald to float down the Isar River to Munich. The beautifully colored houses with ornately carved wooden porches and gables are supposed to date from this period, five hundred or more years ago.

When value in exchange is identified with value, then the project of using nature to the best human advantage (to combat scarcity) becomes identified with a growing economy. People assume that a growing economy helps humanity.

The economy grows when the *monetary* worth of production increases. A subsistence, commons-oriented society produces values primarily for use, not monetary exchange. Due to this dissociation of production from money, subsistence societies impede economic growth (as measured by monetary transactions). They must be replaced for the (exchange) economy to expand.

Such replacement began as early as the fourteenth century in England, where nobles learned that they could sell wool for gold and trade gold for luxuries. Instead of letting peasants use all their land to grow crops for local consumption, they reserved some land for raising sheep. They did this by enclosing portions of the land and expelling peasants from enclosed areas.

Because raising sheep on enclosed portions required fewer workers than did the agriculture that sheep raising replaced, many peasants were thrown out of work. The remaining land was not productive enough to support all of them, so many were dispossessed and without adequate means to meet the material needs of life. As commerce grew, subsistence, commons-oriented ways of life in which use values predominate were reduced.

Dispossessions took place on a grander scale in the newly discovered Western Hemisphere. As with enclosures in Europe, the lure of commerce was the major impetus for exploration that resulted in Columbus's voyages and European conquest and occupation of the Western Hemisphere. Columbus was looking for a trade route to the Far East and gold, the medium of exchange.

We self-servingly refer to Europeans who took up residence in the "New World" as "colonists." When I was going to school, Asians who entered Europe around the time of Genghis Khan were referred to as "invading hordes." Empathizing with those already living in the Americas at the time of Columbus, we may find "invading hordes" an appropriate term for uninvited Europeans.

The European invasion was partly a response to the uprooting of peasants caused by commercial expansion. Dispossessed, unemployed peasants were available for emigration to the New World. So the effects of commercialism—the dispossession and poverty it caused—motivated people to invade the Americas, where they did to its natives exactly what had been done to them. They seized the land, dispossessed its occupants, and enclosed commons areas.

The dispossessions, unemployment, and destitution caused by commercial expansion were what economists today call externalities. These problems and their associated human suffering were not figured into the profit and loss statements of commercial entrepreneurs, because the entrepreneurs did not have to pay for them. They were external to economic or business calculations, just as toxic wastes and their associated health effects are externalities when businesses that generate them do not have to pay for their ill effects.

Using an impious metaphor, one might say that dispossessed peasants ready to invade, conquer, and kill indigenous people around the world were the major toxic wastes of the commercial revolution. They were shipped around the world by governments that wanted to get rid of them, just as some chemical and nuclear wastes are today. Britain not only shipped its own convicts to colonies, such as those in Australia and Maryland, but also accepted for a fee convicts from other countries for shipment to these colonies. Today Britain specializes in handling foreign nuclear waste.

The result of displaced Europeans invading the Americas was not immediately the displacement of indigenous populations who, in turn, went off to displace others. This tragedy was largely replaced by other tragedies—slaughter and disease. The record of Christopher Columbus is instructive. When he first landed on the island he called Española, where Haiti and the Dominican Republic are today, there may have been eight million people living there. Fifty years later official Spanish records put their number at two hundred. This was the result of slavery, inhumane imprisonment, and slaughter, as well as disease and starvation. Besides those on Española, another seven million indigenous people met the same fate *in the Caribbean Basin alone!* Displaced indigenous people did not survive in sufficient numbers to displace anyone else.

Repopulation was accomplished in part by importing slaves from Africa to work on plantations. Population increases among descendents of Europeans, slaves, and indigenous people now make the situation in many places around the world resemble those of an earlier time. Peasants displaced by commercial enterprises are destitute and starving for want of arable land available to them. Central America provides excellent examples. In Costa Rica most peasants had small farm plots in 1820, but forty years later, after coffee plantations came, most had to find wage labor.

The main justification for enclosures, reducing scarcity, is sometimes combined with religion by supposing that God commanded people to subdue the earth to reduce scarcity. A British Columbia Indian Agent wrote in 1880:

> Some of the old Indians still maintain that the lands over which they formerly roamed and hunted are theirs by right. I have to meet this claim by stating that as they have not fulfilled the divine command, "to subdue the earth," their pretensions to ownership, in this respect, are untenable.

In sum, as commerce increases, increasing portions of the earth's surface are removed from use for subsistence.

Third Part Continued: Jeopardy Increases

One result of commercially inspired enclosures designed to manipulate nature efficiently for human benefit is increased human jeopardy. People uprooted from traditional ways of life that provided adequately for their needs are threatened by starvation when they cannot find sufficient wage labor. When increasing portions of the land are held by a decreasing proportion of the population, a concentration of power grows, and with it grows the percentage of people in jeopardy of having their essential needs ignored by those in power. By 1876, for example, six-tenths of one percent of the population of England and Wales owned 98.5 percent of agricultural land.

The logic of using land for its potential to provide exchange values produced similar results in Central America. By the time of our Great Depression most land in four out of five countries was controlled by less than 2 percent of the population. Because land was vital for subsistence, this 2 percent controlled the lives of the other 98 percent. In Guatemala in 1952 2 percent of landowners owned 72 percent of the land, while more than half of the agricultural population owned less than 4 percent.

The control of vital resources by the relatively few people who manage it for commercial advantage leads to widespread hunger. By the turn of the century in El Salvador, formerly self-sufficient peasants, deprived of access to land by coffee plantations, became wandering, hungry laborers looking for work in other countries.

Because the use of fertile land was devoted to a cash export crop, El Salvador had to start importing food. Food was thus increasingly obtained and consumed on a cash basis. Few poor people had enough money to afford a minimally decent diet, much less the varied diet they had formerly enjoyed through cultivation of their own plots. Matters degenerated further in the 1960s and 1970s with the growth of beef exports from Central America. By 1984 malnutrition afflicted 72 percent of all infants in El Salvador.

The story throughout the Third World is the same. Thailand uses its arable land increasingly to grow such export items as poultry, frozen fruit, and especially cassava, which it exports to feed foreign livestock. More than half of its preschoolers are undernourished, and fifty-thousand children die of malnutrition each year, while national income increases.

It should come as no surprise that hunger, malnutrition, and starvation result from taking land away from peasants. Food activists Frances Moore Lappe and Joseph Collins sum it up:

> When the majority of people have been made too poor to buy the food grown on their own country's soil, those who remain in control of productive resources will, not surprisingly, orient their production to more lucrative markets abroad. Fruits, vegetables, coffee, feed grain, sugar, meat, and so on are shipped out of the third world . . .

In Brazil, for example, agricultural exports increased dramatically during the decade of the 1970s while per capita food supplies for the urban population dropped by one-fifth. By the early 1980s hunger's toll had reached at least half of Brazil's population, up from one-third two decades earlier.

People who find a place in the commercial economy have reason to feel jeopardized too. As they depend increasingly on exchange with others to meet their essential needs, they are increasingly vulnerable to disruptions in supplies, blockages of trade routes, devaluations of currencies, and so forth.

In addition, those who are uprooted from subsistence ways of life and fail to find a niche in commercial society threaten commercial links through robbery, murder, and sabotage. The dispossessed have been a major law and order problem since the sixteenth century and have been active in periodic peasant revolts. One began in Chiapas, Mexico, in January, 1994.

Fourth Part: Centralize Power to Reduce Jeopardy

Insecurities associated with commerce justify increased centralization of political power. Commerce thrives where there is political stability, where robbers are caught and punished, where there are good roads that are maintained uniformly, and where there is a common currency that can be used as a medium of exchange. It is no accident that modern nation-states are responsible for maintaining these conditions, and that they grew with commercial expansion that ended the Middle Ages. This is when Spain and Portugal, for example, developed into roughly the states that we know today.

At present these forces continue to operate. The European Community is attempting to improve conditions for trade such as free and easy passage across borders and a common currency in Europe. The North American Free Trade Agreement (NAFTA)is designed to stimulate commerce by removing barriers to trade within North America and the General Agreement on Tariffs and Trade (GATT) is designed to facilitate commerce worldwide. In each case, power is relinquished at a relatively local level (today it is at the level of the nation-state) and given to a power that is exercised over a larger geographic area.

The concentration of political power in the contemporary world can be seen also in the increasing activism of the United Nations' Security Council. With greater frequency it authorizes the use of force to address international difficulties.

Fifth Part: Address Danger with Oppression

Power concentrations are used to oppress people considered dangerous. This is evident in the Third World where repressive dictatorships suppress popular revolts. Governments in more affluent countries are less repressive because affluence makes the populace more cooperative. But this affluence is largely responsible for poverty, revolt, and repression in the Third World.

The system resembles a pyramid scheme. In such schemes, an individual (B) is supposed to gain by sending some money to one person (A) as part of a system organized to require several others (C,D,E,F, and G) to send money to (B). Of course, the others seldom send the money, so instead of making money B loses her initial contribution, which is a flaw in performance. But there is a flaw in theory as well. Even if the others were to send money, people eventually would lose their investment simply because there would be no more people left to join the scheme. With participation increasing geometrically (each new participant is supposed to recruit five others), the whole planet would be quickly absorbed. The largest number of people would lose because they contributed to someone, but there would no longer be people left outside the scheme to contribute to them.

Land use is the same. People in Europe who were dispossessed could move to the colonies and dispossess others. Except where these others were exterminated, which happened all too often, newly displaced people were in a bind because the whole planet was quickly absorbed. There were no places left for them to go; no others left for them to dispossess. Just as people at the beginning of a pyramid scheme may become rich, the people who initiated commercialism became rich. But just as the riches of the pyramid scheme are paid for by many losers, so are the riches of commercially successful people and nations. Today, poverty-stricken people in the Third World are among the losers.

Because the social pressure of discontent in the Third World cannot be relieved by conquering, colonizing, and exploiting other lands, it must be contained by force; hence the prevalance of repressive military dictatorships. Affluent countries generally support this repression. They established and aided the repressive regimes of Batista in Cuba, the Shah in Iran, Marcos in the Phillipines, and Pinochet in Chile, to name but a few. The United States helped these dictators by giving them favorable conditions of trade, lending them money, training their military, selling them arms, and foiling their potential rivals.

Whatever reason may be given, such as support of freedom or

opposition to communism, the guiding thread was clearly benefit for commercial interests. The United States supported repressive regimes whether or not there was a credible threat of communism and whether or not the alternative to them was democracy, because these regimes benefitted commercial interests by protecting property rights and keeping taxes low.

People in affluent countries are apt to think that protecting property rights and keeping taxes low are good for everyone in society. Turn-of-the-century French satirist Anatole France had a reply. He noted that in France the law is majestic in its equality; the rich and poor are equally forbidden to sleep under railway tressels. Just as only the rich can realistically benefit from unimpeded rail traffic, and only the poor from using tressels to keep out of the rain, only propertied people can benefit from the protection of property and low taxes. Poor people are more likely to benefit from social programs that higher taxes might fund.

The most commercially successful countries, such as the United States, have overthrown, or helped to overthrow, regimes in poor countries that promised genuinely to help the poor. The United States helped to overthrow democratically elected governments in Guatemala and Chile because these governments were going to raise taxes and distribute land to peasants. This would have enabled many peasants to be self-sufficient and lead tolerable lives but at the same time would have deprived commercial interests of money and some access to the country's natural resources.

(I sometimes talk to students about capital punishment. They often emphasize the danger of allowing repeat killers to live. This reminds me of the U.S. role in the bloody coup that deposed a democratically elected regime in Chile. I tell my students that they may be right. Henry Kissinger may kill again.)

Fifth Part Continued: The Oppression of Women

Women especially are disadvantaged by commercialism. While raising three children in a lower-middle-class suburb, my mother always said she did not work. An intelligent woman of my own generation told me about ten years ago that her financially successful husband owns the house, cars, and all other material possessions used by the family. He earned the money; she was just raising their four children.

When value means "exchange value," many contributions by women are ignored because they are not done in exchange for money. Worldwide, women do most of the unpaid gardening, child-rearing, and housework for household consumption. In commercial society, this is not considered work because there is no pay.

The misperception that women work less and are less productive than men is used to justify economic discrimination against them. In many parts of the world women are not allowed by tradition to own land in their own names, but they had traditional rights to use commons areas for food production. As food production is increasingly commercialized, and commons areas decline, women are deprived of these traditional rights.

As women lose traditional rights men gain additional power over them, and often ignore women's potentials and needs. Worldwide, women constitute an increasing percentage of those who are illiterate, because educational opportunities are directed toward men. In India, food also is directed toward men. Compared with men, per capita food intake and life expectancy among women have declined, and women constitute a decreasing percentage of the population.

In sum, just as commercialism increases the gap in life chances between rich and poor, commercial and indigenous, First World and Third World, so it increases the gap between men and women.

Comparative Advantage and the Promised Future

Many people in affluent countries are unaware that their commercial economies hurt people in the Third World. We have been told that free enterprise helps everyone. If each country concentrates on what it can produce more efficiently than anyone else and then trades with other countries for everything else it needs or wants, the sum total of wealth in the world is maximized. We are told that this eventually helps everyone. What goes around comes around. This is the theory of comparative advantage.

According to this theory, poor countries in particular are helped by producing increasingly for export, as this gives them foreign exchange earnings that can be used to buy needed capital. As wealthy countries become richer, they can afford to buy more from the poor. So the best thing that wealthy countries can do to help the poor is to increase their own wealth. What luck! By happy coincidence, increasing their own wealth is just what rich nations had in mind.

This convenient theory is just a global version of trickle down economics. If the rich get richer, everyone will benefit in the long run from purchases that rich people make. One reason for skepticism is that the rich got significantly richer in the United States in the 1980s without benefit to the poor. The same happened in Great Britain, according to British economists Tim Lang and Colin Hines. "Wages for the top ten per cent of male full-time workers gained 35 per cent between 1979 and 1991, whereas for the bottom tenth they went up only 5 per cent. In the same time the taxes for the poorest fifth went up by 6 per cent of their income." If trickle down economics does not work domestically, why should we think it will work internationally?

Direct evidence is mounting that trickle down economics has not been working on a global scale. In the 18th century, people in Europe were about twice as wealthy as people in India. In 1965 the World Bank estimated the income gap between people in rich and poor nations to be 15 to 1. By 1990 it had risen to 33 to 1.

There is little prospect of reversing this trend. Currently, even after all foreign aid is included, the net flow of money from poor countries to rich is $43 billion per year for "interest repayments, repatriated profits, capital flight, royalties, fees for patents and information services. . . . "

> Developed economies appear to have all the trump cards in their hands—capital, technology, control of communications, surplus foodstuffs, powerful multinational companies—and, if anything, their advantages are growing because technology is eroding the value of labor and materials, the chief assets of developing countries.

In light of these figures, it is no surprise that low-income countries have fallen deeper and deeper into debt. Yale historian Paul Kennedy writes, "Only a third of the $1,200 billion debt the Third World owed the banks of the First World in 1990 was the original debt. The rest was accrued interest and capital liabilities." The need to make debt payments forces poor countries to curtail basic services to their citizens. This applies equally to democracies, such as Mexico, as to other countries. Third World countries with significant foreign debts try to repay debts in hard currency (the currencies of lending nations and institutions) by maximizing the use of their natural resources to produce commodities for foreign exchange. This means uprooting more peasants and ignoring local needs to supply the demands of foreign consumers.

One result is a flood of indigent people toward cities. Mexico City, for example, now has a population of twenty-five million. These people have nothing like the ability of traditional farmers or pastoralists to care for themselves. Their lives are sustained by makeshift adaptations to the available refuse of the commercial economy. This is not just a metaphor. In Mexico City many people live literally on garbage heaps where life expectancy is not long. Brazil is infamous for charges that its street children are simply murdered.

Child prostitution increases yearly at alarming rates in many poor countries, notes writer Aaron Sachs.

> Brazil alone has between 250,000 and 500,000 children involved in the sex trade, and a recent study conducted by the Bogotá Chamber of Commerce concluded that the number of child prostitutes in the Colombian capital had nearly trebled over the past three years. Similar increases have occurred in countries as geographically and culturally disparate as Russia and Benin. But the center of the child sex industry is in Asia: children's advocacy groups assert that there are about 60,000 child prostitutes in the Philippines, about 400,000 in India, and about 800,000 in Thailand.

Most are girls under the age of sixteen.

Hunger continues to be a problem in poor countries. Twelve to 20 percent of the world's population, somewhere between 700 and 1,200 million people, are seriously malnourished. Perhaps a single statistic is sufficient to dramatize the effects of this malnutrition and the failure *so far* of commercialism to help poor people: *40,000 children die everyday from inadequate diet.* Weakened by malnutrition, they usually die of preventable diseases, often of dehydration caused by diarrhea produced by contaminated water. These are just children, and this is *everyday,* all around the world, not just in areas celebrated for drought. Many of the malnourished children live in Central America where bananas and beef are grown in great quantities for export to the United States. The reputed benefits of commercialism have so far eluded these people.

But what about the long run? Maybe we are just in a period of painful transition that will result in such great benefits later as to have made all the suffering worthwhile. This is another convenient theory when advanced by people who are excluded from the present "necessary suffering." It is like rewards in the hereafter in some religious traditions. Put forward by people in advantageous positions, the promise of supernatural rewards helped to mollify the suffering population and protect the status quo. Karl Marx rejected such religion as an opiate of the masses. Faith in the economic future is no better. Current evidence suggests commercialism exemplifies the five-part pattern. Human oppression results from efforts to help humanity.

Economic optimism is supported by success stories in such emerging industrial centers as South Korea, Taiwan, and Hong Kong, where comparative advantage appears to be working. I discuss industrialism next to show that optimism rests on the fallacy of pyramid schemes.

Chapter 3

Industrialism

Standardization and Centralization

We departed Mittenwald for Garmisch-Partenkirchen, which *Fodor's* considers the Alpine capital of Bavaria. The 1936 Winter Olympics were held there.

The center of Garmisch shows an American influence—McDonald's, the paradigm of industrial standardization. Their products are standardized throughout the world; their production methods are industrial in a field, food preparation, that is otherwise largely private and individualized; they are a large corporation with significant capital investment; and they are centrally organized and controlled. Of these traits, standardization clashes most with my vacation ideal of unique experiences in a foreign country.

I cannot be entirely negative about standardization. Without it, manufacture as we know it, including the plane we took to Europe, would be impossible. The plane is constructed from parts that were designed and made to standard specifications for planes of this general type, not just for this one plane in particular. Making parts for many planes at the same time saves money. Safety is improved when planes are built according to a standard design tested by experience. A unique, made-to-order plane may or may not work in the first place, or may work for just one hundred flights. No one would have the data needed to predict its endurance.

Industrial production ("industrialism" for short) typically involves, besides standardization, mass production, division of labor, centralized management, high capital costs, and relatively high energy use. The division of labor facilitates mass production because people generally can produce more per hour when they repeat a single task. Repetition makes people expert and facile in their work, and minimizes time used switching from one task to another.

Standardization is needed for mass production when labor is divided among many people. The product is usually a complex whole of interrelated parts. Standardization is needed to ensure that parts made by different people interrelate properly.

Centralized management coordinates the activities of specialized workers. Because there is often an economy of scale in production (production of large quantities reduces the cost per unit), relatively large organizations are typical. This makes coordination particularly important.

It takes a lot of money to start large organizations, so industrialism is relatively capital-intensive. Significant monetary investment precedes any financial return.

Finally, industrial production tends to use more energy than preindustrial production. This energy comes mostly from falling water, fossil fuels, and nuclear fission. Generating power contributes to high capital costs and economies of scale.

The goal of all this is increased human productivity in the service of humanity. If scarcity is endemic to the human condition, as mainstream economists maintain, efficiencies resulting from standardization, mass production, divisions of labor, and centralized management can all contribute to improving the human condition by reducing scarcity.

My trip with Patricia was a case in point. We had limited time and money and were concerned about our physical safety. At the same time, we believed we could benefit from the trip. The major elements of industrialism combined to make possible our quick, relatively safe, and affordable flight from Springfield, Illinois, to

Munich, Germany. Endemic scarcity was reduced because we *wanted* to go. If, as modern philosophy and mainstream economics tend to agree, there is no difference between wants and needs, the frustration of our desire would have indicated scarcity.

Industrialism and Commercialism

Industrialism is intimately related to commerce. The justification of reducing scarcity applies as much to industrialism as to commercialism. Commerce also supplies the profit motive for industry to lower costs of production and increase the number and variety of things produced for exchange.

Industry helps commerce in a number of ways. Some industrially produced items improve the means of engaging in commerce. Steamships, railroads, and trucks, for example, help transport goods to market, and refrigerators enable perishable items to survive the journey.

Commerce is fostered also by whatever increases the efficient operation of large organizations. Industrially produced communication technologies—telegraph, telephone, computer, and satellite—do this. Radio and television also help inform the public of new "needs" and new products to meet those needs.

Because commerce and industry are mutually reinforcing, they equally exemplify the five-part pattern. Both distinguish people from nature, consider people to be jeopardized by endemic scarcity, and justify exercising increasing power over nature to reduce scarcity. Unfortunately, as with commercialism, industrialism creates new jeopardy, which justifies increasing concentration of power, and this power is ultimately used to oppress people. It is bad news, but here it is.

The Industrial Revolution, Colonialism, and Slavery

The industrial revolution began in England in the 1780s. It spread to the continent of Europe early in the nineteenth century and to the United States later in that century. In view of its mutually reinforcing relationship with commerce, and the tendency of commerce to foster colonial oppression, it is no surprise that such oppression increased. Britain was the greatest industrial power and had the largest overseas empire. The colonial empires of France, Germany, Japan, and the United States grew with their industry.

Industry and colonial conquests are related partly because industries produce technologically innovative weapons in great quantities, from repeating rifles in the nineteenth century to "smart bombs" in the twentieth. This facilitates the military domination of nonindustrialized people. Developments in shipping and aeronautics help transport troops and weapons to remote locations. Industry made overseas conquests easier.

Industry also made conquest more attractive. Industrial manufacture enables people to make things more quickly and cheaply than before. The Industrial Revolution began in the textile industry when methods were discovered for spinning and cleaning cotton on a mass scale; thus, turning cotton-growing countries such as India into colonial outposts seemed like a natural progression of events.

Industrialization also fosters colonial conquest because increased output requires additional raw materials. A country that may have enough forest to supply preindustrial needs for wood for home building may need additional lumber when mass producing prefabricated housing for export. This is an incentive to conquer, or at least control, another country's forests.

The need for markets for industrially manufactured goods is another incentive. When all countries produce their housing domestically no country is trying to control housing consumption overseas, but when a country specializes in manufacturing prefabricated housing, its workers' job security depends on continued consumption of its products by people in other countries. Domestic politics requires attempts to control that consumption.

While prefabricated housing has not reached that point yet, this kind of issue is at the heart of recurrent trade tensions between the United States and Japan. The United States tries to gain greater access to Japanese markets. The problem is endemic to industrialism because quick and easy massproduction means there is always a risk of oversupply. In the very first industry, British textiles, Britain's incentives to control India included not only the importation of raw materials, but the export of manufactured goods.

With more goods coursing around the world, and domestic job security riding on their safe delivery, industrial countries needed to improve and secure trade routes. Trade routes were improved by creating the Suez and Panama Canals. Securing access to canal zones motivated the United States to control Central America and Britain and France to control Egypt.

In general, industrialization increases commercial interaction around the world, as any one group of people can produce something in greater quantity and for a larger market. Interaction leads to interdependency; people in one part of the world depend on people in other parts for raw materials, safe passage, and lucrative markets, making each vulnerable to the others. People try to reduce their vulnerability by controlling those whose behaviors can hurt them, so as industrialization increased in Europe, the United States, and Japan, so did control over colonies. The result was the increased oppression of colonial people.

Another effect of industrialization was the promotion of slavery in the American South. The Industrial Revolution began in England with the mechanical spinning and cleaning of cotton. In 1765 only half a million pounds of cotton were spun by hand, but in 1784 twelve million pounds were spun by machine. Steam engines were then introduced to power the spinning machines. The efficiency of finishing cotton for human use improved dramatically for several decades.

The increased demand for cotton was met partly by the development of cotton plantations in the New South—western Tennessee, Alabama, Mississippi, and Louisiana—whose population tripled between 1810 and 1830. Slavery increased as well. Historian Paul Johnson writes:

> For every 100 acres under cotton in the New South, you needed at least 10 or perhaps as many as 20 slaves. The Old South [Virginia, the Carolinas, and Georgia] was unsuited to cotton but its plantations could, and did, breed slaves in growing numbers. Slave breeding now

became the chief source of revenue for many of the old tobacco plantations. . . .

One indication of the increased interest in, and value of, slavery was the increased price of slaves. According to Johnson, "Before the cotton boom, the price of slaves in the United States had been falling: In the quarter century 1775–1800, it slid by 50 percent. Now it began to rise. In the half century 1800–1850, it rose, in real terms, from about $50 to $800–$1000." Johnson concludes:

> [I]t was the early 19th Century world cotton boom which created "the South" as we understand it, a geographic entity united by its defense of slavery. . . . The notion that Southern slavery was an old-fashioned institution, a hangover from the past like serfdom in Russia, was wrong. It was born of the Industrial Revolution, high technology and the commercial spirit catering for mass markets of hundreds of millions of consumers.

Faith in Progress

Some optimistic thinkers maintain that slavery and colonial oppression are temporary phenomena. Consider slavery first. Optimists claim that slaves are economically advantageous only when performing labor-intensive work, as on cotton plantations. Industrialism improves efficiency largely by replacing human labor with machines, thereby depriving slavery of its economic rationale.

This reasoning fails to note the cause of resurgent slavery in the American South. Industrialism had, through the introduction of machine efficiency, replaced human workers or made human labor much more efficient, in *some* aspects of a production process (cotton spinning and then cotton cleaning). This created the need for more workers at low (or no) wages in other parts of the production cycle (cotton growing). Slavery increased because breakthroughs in production were not simultaneous throughout the production cycle.

This can be expected to occur often, since it would be sheer accident if technological, industrial breakthroughs were to take place and be applied simultaneously in all phases of production in an industry. Because breakthroughs are seldom simultaneous, industry continues to need cheap labor. After slavery was ruled out politically in Britain and the United States, the Industrial Revolution generated unskilled factory jobs appropriate for slaves. As things turned out, working life for laborers of the time was little, if any, better than slavery.

The same thing exists today in the textile industry. Cloth can be manufactured so efficiently that pressure is put on phases of clothing manufacture that cannot be fully automated, such as sewing. Sweatshops are reported to exist today in the United States just as they existed near the turn of the century. According to one report, there are twenty thousand of them in New York City alone. Most workers are recent, often illegal, immigrants and are treated little better than slaves.

Matters appear to be worse around the world. A National Public Radio Report in March 1993 contended that there are currently two hundred million slaves worldwide! In absolute numbers, then, there may be more slaves in the world today than at any other time. This does not count people who are compensated no better than slaves that are working in the United States and other countries for below-subsistence wages. (Many of these people live in the United States since the minimum wage is no longer sufficient for subsistence). In sum, slavery does not tend to disappear peacefully through economic forces in industrial societies.

Consider now the optimistic forecast that sees the oppression of Third World people as temporary. Optimists maintain that when Third World countries become industrial powers in their own right, they will obtain the liberating benefits of material prosperity, just like people did in Europe, the United States, and Japan. Taiwan, Korea, and other Pacific Rim countries are obtaining these benefits already. In time, all humanity will benefit.

This reasoning suffers from the same fallacy of pyramid schemes noted in connection with commercialism. Where manufacture of material goods is concerned, limited global resources prevent everyone from doing what industrial countries have done. According to the authors of *Beyond the Limits*, "Human use of many essential resources and generation of many kinds of pollutants have already surpassed rates that are physically sustainable. Without significant reductions in material and energy flows, there will be in the coming decades an uncontrolled decline in per capita food output, energy use, and industrial production." As a result, "Poverty cannot be ended by indefinite material growth; it will have to be addressed while the material human economy contracts."

According to the 1991 World Bank Development Report, material standards of living declined during the 1980s in forty countries, where eight hundred million people live. Such decline will engulf the world unless we change direction.

It could hardly be otherwise. Each American uses twenty or thirty times the resources of people in many populous countries in Asia, Africa, and Latin America. If problems of pollution and global warming exist now, imagine what they would be if the present level of industrial activity increased twenty or thirty times, which would be needed to permit others in the world to enjoy the American standard of living wastefully. There was general agreement at the 1992 World Summit on the Environment that this is impractical.

Like the bad news about slavery, this message is hard to accept because it goes against our culture's faith in progress. I can only suggest keeping in mind that in secular matters faith should be tempered by evidence. Whether the issue is slavery or the concentration of industry's benefits among rich nations, overwhelming secular evidence now indicates that industrialism does not lead to a more humane world.

Class Stratification

Besides the growth of slavery and environmental limits to the worldwide expansion of industry's benefits, a third problem belies faith in progress. Industrial production tends to oppress people *within*, as well as outside, industrial countries.

People are oppressed along with nature due partly to economies of scale, which favor production in large organizations that require considerable monetary commitment before they yield income. This gives those who control financial capital great power over others. Financiers decide who begins which industries where. Workers who depend on industry for their livelihood may be required to make wage concessions to avoid relocation of their industry to the Third World where labor is cheaper. Owners and managers claim that competition forces them to offer low wages that minimize production costs.

Owners and managers are taught that humanity is served best when governments support private enterprise, and all agents in the economic arena act selfishly. In the words of Robert L. Bartley, editor of *The Wall Street Journal,* "[T]he fundamentals of human nature are universal. . . . [H]istory suggests that economic development depends on harnessing the acquisitive instinct—greed, if you must, or as Adam Smith put it, self-love." Thus, pure selfishness is reasonable, at least in business. Under these conditions wages sink to subsistence or less unless governments intervene. Massive poverty accompanied the Industrial Revolution in England and France when government intervention was weak.

Through the control of government by ordinary people, representative democracy is supposed to produce government policies that protect workers from exploitation. But the benefits of democracy are muted by the disproportionate political influence of wealthy people. For example, money plays a larger part in U.S. elections than in those of any other industrialized country, while the gap between rich and poor is greatest and the income tax paid by the rich is lowest. Compared with other industrialized countries, U.S. law provides workers with the least protections against unemployment, the worst guarantees of access to health care, and the lowest minimum wage (relative to the cost of living). Many full-time workers in the United States live in poverty and many are even homeless because wages are so low. We are told that competition disallows raising the minimum wage.

Competition is used increasingly also to justify exposing work-ers and the public to environmental hazards. Free trade is supposed to maximize the benefits of competition by fostering fair competition worldwide. But GATT and NAFTA already have been used to require environmental deregulation in the name of fair trade. According to a GATT panel, "[N]o country may use trade restrictions of any kind to enforce public-health laws unless that country can prove its policy was the least trade-restrictive."

Many U.S. laws restricting the use of such chemicals as DDT and heptachlor, for example, could be invalidated under GATT through international adoption of weak environmental standards. Because standard-setting is not done democratically, international agreements such as GATT and NAFTA enable industry to avoid the confrontation with people's needs that democracy is supposed to require. It is worth remembering that Dickens's England was the result of industrial production without adequate governmental safeguards for average people. In sum, competitive, efficient mass-production tends to disadvantage many people, even though it is supposed to maximize the benefits that people glean from nature.

Skepticism about Darwin's Theory

Why do people stand for this? One reason, I believe, is that competition, and resulting problems for losers, are widely viewed as natural and inevitable. They are basic to evolutionary processes responsible for our very existence.

Frankly, I am skeptical. I have no doubt that evolution occurred. Over millions of years, current species evolved from other species. But how and why did this occur? I think now that no one really knows, although I used to assume that Darwin's theory, which emphasizes competition, was essentially correct.

Today's Darwinian theorists maintain that species evolve principally when genetic mutations cause offspring to differ from parents, and the differences confer competitive advantage. Advantageous differences are those enabling individuals to make better use of available resources to produce more progeny able to do the same. Because the advantageous differences are genetically based, they can be passed on to succeeding generations through procreation. Because individuals with this genetic make-up are more successful than others in obtaining and using vital, scarce resources, they and their progeny become the new norm for the species. The species differs from what it used to be because only the fittest and their progeny survive. This is fitness maximization.

One problem with this theory is that a basic premise, that competition alone guides natural selection, remains only an article of faith. It is impossible to know if species maximize fitness because any change in a species will be helpful or not depending in large part on what the species was already like. For example, genetic mutations resulting in a new gill slit may help some kinds of fish, but are unlikely to help dogs, given the way dogs have already evolved. These preexisting conditions that affect the utility of a given change are called constraints.

Unfortunately, not all constraints are easily identified. Living things interact with one another, and with inorganic nature, in ways we do not understand, so we cannot know what constraints affect the utility of many possible changes in a species' make-up or habits, much less know all the constraints species were subject to during the course of evolution. So when we see that a species evolved in a certain way, we have, in principle, no way of knowing whether the evolutionary change was fitness-maximizing, given the species' constraints.

The situation is like this: Suppose you are told that someone is generous, she always gives all she can to charity. She is charity-maximizing, but, of course, she has many financial obligations and a limited income. These are her constraints. Suppose that you have an incomplete knowledge of these constraints. Even if you knew exactly how much she gave to charity you would have no way of knowing if she was really charity-*maximizing,* as her financial obligations and income may have enabled her to give more. Maybe some other principle also guides her behavior, such as love of expensive chocolate. You simply cannot know from observation of her charitable contributions alone.

Similarly, a record of how a species evolved under constraints does not enable us to know if that development was fitness-*maximizing*, because we do not know what the constraints were. Random mutations may have allowed the species to evolve in ways that would have made its members even more competitive, but there is some other principle guiding evolution in addition to competition for available resources to produce similarly competitive progeny. We simply cannot know. So the key contention that competition *alone* selects biological traits for survival can never be proved.

Furthermore, some species' methods of procreation seem unlikely to be the most competitive possible. Why do salmon, who live most of their lives in the ocean, swim far upstream in fresh water to spawn? It would seem much simpler to evolve the means of spawning in the ocean. Perhaps, given the constraints present when relevant mutations occurred, this is the most competitive the species could be. On the other hand, maybe not. We cannot know.

Another puzzle concerns sexual differentiation. Why does it exist? Individuals pass on traits through transmission of their genes. The first living things reproduced asexually, such as by dividing. This is like cloning. If the point of reproduction is to produce progeny like oneself, as evolutionary biologists maintain, why would beings who could do this perfectly (by cloning) evolve to reproduce sexually. In sexual reproduction each partner gets to pass on to progeny only one-half of his and her genes.

Sexual reproduction may be good for the species in the long run. It makes progeny genetically unique (except for identical twins). This may enable a species to survive environmental changes, because some of its unique individuals may by chance be adapted to the new conditions. An asexual species of more uniform individuals may be wiped out completely.

But evolutionary theory cannot use this advantage to explain the development of sexual reproduction. According to that theory evolution is not directed toward long-term ends by some spirit, force, or mind. It is influenced only by competition among contemporaries, and asexually reproducing species are not known for competitive strategy or foresight. Again, it is not clear that competition is the only principle of biological selection. Other puzzles have not yet been solved by evolutionary theorists. How could altruism, helping others, ever have started in competitive evolutionary contexts? Also, why does the fossil record suggest rapid, rather than more gradual, changes in species, and how could all genetic changes needed for a complex organ such as the eye accumulate if each occurred independently (genetic mutations are random events) and conferred no competitive advantage on first appearance?

A final problem, among the many that could be considered here, concerns the application of nonhuman evolutionary theory to human beings. Human culture powerfully influences human reproduction. People almost everywhere help their kin to survive and reproduce. But they define kinship socially, not biologically, so they often help people who are biologically unrelated. In several cultures people practice infanticide and then adopt the children of slain enemies. This seems completely opposed to the idea that evolution is a competitive struggle of individuals using resources maximally to produce maximal offspring who will do the same.

I leave these, and many other puzzles, to experts, concentrating instead on two considerations. In outline they are: First, there is a remarkable similarity between Darwin's account and industrial processes. Second, Darwin's theory is often invoked to justify inequality between rich and poor. These two considerations give social explanations for the scientifically unjustified faith that competition is the master principal of life, including human social life.

The Industrial Evolutionary Theory

Consider the similarities between industrial production and life as seen by Darwin. For Darwin, reproduction is essentially the transformation of the earth's resources into standard copies of an original model. In reproduction the producer is herself the original model, otherwise reproduction resembles industrial mass production of standardized items.

Industrial producers are supposed to compete against one another. Those who can produce the greatest quantities most efficiently are successful. Their products are bought and they "survive" to produce more. Similarly, those who reproduce most efficiently are successful in biological competition, and their offspring survive to produce more.

Progress is made in industry when manufacturers modify products to attract more consumers, and/or make production more efficient. Such changes become widely imitated due to competitive pressure. In biological evolution, chance imperfections in the copying process occasionally produce nonstandard individuals who can garner more from the environment, better escape preditors, or for some other reason reproduce more efficiently than the standard model. Through competition these nonstandard individuals become the new standard. As evolution is popularly perceived, this constitutes an evolutionary advance.

In industry products disappear or become marginal when ill-suited to newly evolved industrial conditions. The buggy whip is the classic example. In biological evolution, species become extinct or marginal when ill-suited to newly evolved biological conditions. Dinosaurs are the classic example.

In manufacturing, efficiency is often achieved through job specialization, as on the assembly line established by Henry Ford. The analogue in evolutionary theory is species' specialization. For example, members of a given species are efficient predators of only certain foods among those available. Such specialization helps members of different species coexist noncompetitively in the same geographic area. They use different resources.

As the diversity of species increases with specialization, species provide food and shelter for one another. So just as in the commercial world specialized people are suppliers to, and customers of, those who have different specialties, which increases total economic activity, biological specialization fosters symbiotic relationships that increase the total amount of life the land can support. Evolutionary biologists sometimes refer to a species' specialization as its "profession."

The idea of progress plays similar roles in industrial ideology and in popular accounts of evolutionary theory. In these views, progress consists in the increased number and variety of beings produced (industrially or biologically). Increases are due to innovations that enable industries and species to use resources more efficiently, to tap resources previously unused, and to become (symbiotically) resources for one another. Tendencies in these directions are natural and welcome, whereas opposite tendencies, destruction of species' habitats and destruction of good business environments, are decried and resisted.

In light of all these parallels, it seems that evolutionary theory is biology viewed through industrial glasses. People result from evolutionary processes, but the theory that explains how and why this biological evolution took place is, in large part, a reflection of commercial and industrial processes.

Indeed, before Darwin published his views concerning nonhuman species, Herbert Spencer articulated a similar position, relating human social and economic competition directly to human evolution. Spencer's work highlights the close connection between evolutionary theory and commercial-industrial processes.

It would be something of a coincidence (possible but improbable) if the truth about human origins just happens to correspond to the way our industrial society operates. I know that people have a tendency to project aspects of their own experience or society onto reality at large. For example, Hobbes, living at a time of civil war, assumed that war was the natural human condition. Aristotle assumed that the moral virtues celebrated in Greece at his time were proper for all people at all times. Most societies use the same word for members of their society and for human beings in general, as if their society typifies humanity at large.

Darwin's theory of evolution seems to reflect this tendency to project local conditions, competitive industrial production in this case, onto a wider reality. It would be surprising if such projection turned out to be scientifically accurate in this case, even though it is wrong in the others.

These considerations suggest that evolutionary theory is probably incomplete. It probably excludes important aspects of reality that are not analogous to industrial manufacture, just as Hobbes's materialist view of reality caused him to miss important aspects of human nature. In particular, competition for resources may not be the only principal of natural selection.

Social Darwinism's Justification of Inequality

Another reason to question the prevailing theory of evolution is the role this theory plays in justifying inequality. The Industrial Revolution produced extremes of wealth and poverty, attributable in part to the selfishness of rich people. Social Darwinists defend such selfishness as biologically inherent. In evolution's competitive struggle for existence, selfish individuals would defeat altruists and leave more progeny with inherited selfish tendencies. So contemporary people are naturally selfish. This is probably the basis for *The Wall Street Journal*'s editorial that greed is a fundamental universal of human nature.

When evolutionary theory is interpreted in this controversial way, it implies that genuine altruism is impossible. Because it is senseless to expect people to do the impossible, it is senseless to expect owners, entrepreneurs, labor leaders, politicians, or anyone else, to be altruistic.

Many Social Darwinists go further, maintaining that selfishness is good because it promotes progress. Wealth stems from improvements in productivity that are part of the overall evolutionary advance of humanity in its struggle with nature. So the wealthy are benefactors of humanity and should be encouraged in their selfishness. Restraints on rich people, who represent the evolutionary avant-garde, hurt humanity more than they help because they penalize and discourage society's most productive citizens, (as if a person raising a child is less productive than one marketing pet rocks). President Clinton's 1993 proposal to raise the income tax rate of wealthy people was derided by many conservatives as punishment for productivity.

Using an evolutionary model similar to Darwin's, Herbert Spencer claimed in this connection that trying to help poor people is futile because their poverty shows them to be inherently inferior competitors. Many current economists conclude that the only hope for the poor is general economic expansion, as a rising tide lifts all boats. Here is how that is supposed to work.

As already noted, evolutionary theory holds that competition among individuals and species results in species specialization and proliferation that enable natural resources to support more life. Tree species evolve that can live in the shade of other trees, thus expanding the tree coverage of the forest floor, and animals evolve that live in these trees and eat their fruit. Parasites evolve to live in the intestines of these animals, aiding their hosts' digestion while taking a cut of the nutritional action.

The social analogue would be free competition among selfish producers, resulting in the proliferation of increasingly specialized goods and services. This proliferation creates new economic niches, fosters new economic symbioses, and allows a greater number and variety of people to be economically productive. Power mowers enable people to cut their lawns while exercising little. This creates greater demand for health clubs where people can exercise. Air conditioning reduces neighborly encounters on outside porches in hot weather, making health clubs attractive also as places to meet people. Health clubs foster the development of variety and style in sportswear. Jobs are created for poor people, such as in lawnmower manufacture and service, sportswear manufacture and sales, and health club construction and management.

In sum, the best way to help everyone, including poor people, is to promote an expanding economy by allowing selfish rich people to get richer under conditions of competitive free enterprise. This, of course, is just the trickle down theory. It may persist, despite its failures in performance, because it coheres with an attractive Darwinian perspective on human social life. Or perhaps the Darwinian perspective is popular because it supports trickle down economics favored by the rich. In either case, the prevailing competition-oriented acount of how and why biological evolution took place is used to justify gaps in income and wealth between rich and poor.

This account was first propounded as democratization increased in England. Representative democracy places considerable power in the hands of average citizens, who can use it to curb the privileges of wealth and reduce the deprivations of poverty. Redistributions to the poor can be minimized by convincing average people that disparities between rich and poor are justified. The theory of evolution provides such a justification. It is suspiciously like intellectual-scientific cheerleading for society's unequal status quo. Darwinian evolutionary theory may facilitate the human oppression associated with industrialism by helping to legitimate extremes of wealth and poverty that contribute to that oppression.

It also facilitated the oppression of indigenous people. Many nineteenth century evolutionary theorists assumed tribal societies to be cases of evolutionary arrested development. This justified colonial domination, because if evolutionary progress is good "more evolved" societies are better than "primitive" ones. Subjugating "savages" was a duty because it furthered evolution, which helped everyone in the long run. This rationale for oppressing indigenous people reinforces religious, commercial, and industrial reasons.

Sociobiology and the Subordination of Women

In a form known as sociobiology, Darwinian evolutionary theory is currently being used to justify the continued subordination of women. Reproduction, on this view, is really what life is all about. Women have fewer opportunities than men to reproduce. According to Professor of Behavioral Sciences Daniel G. Freedman:

> A human male overproduces sperm (male gametes) by a factor of ten thousand and potentially could populate the universe with his own offspring. . . . Human females, by contrast, . . . are theoretically lim-

ited to some 60 offspring. Given the vast difference in reproductive potential, and if the point of life is to actualize such potential, . . . [m]ight nature not have arranged it so that men are ready to fecundate almost any female and that selectivity of mates has become the female prerogative?

Freedman concludes that human beings are predisposed toward a double standard in matters of sex. Because discrimination against female promiscuity is biologically inevitable, it is moral.

According to some sociobiologists, other areas of justified discrimination against women concern childrearing and employment opportunities. Since the reproductive potential of women is so much less than that of men, women have a larger biological stake than men in each offspring. In order to perpetuate their genes, women are more inclined by nature than men to care for children, and children have evolved to respond best to maternal care. Thus, women naturally want to be full-time mothers. Job discrimination, and other social practices that discourage mothers from working outside the home, benefit everyone.

This view furthers the social subordination of women. It is often championed in the name of family values by people who call themselves conservatives. But American conservatives generally support free market economics, which uses money to measure social contributions. Relegating women to unpaid work stigmatizes them as unproductive in a production-oriented society.

The foregoing account is, I know, one-sided. Some thinkers have used evolutionary theory to promote altruism, cooperation, respect for indigenous people, and equality for women. However, their views have to date had less impact than the Social Darwinist and sociobiological theories just reviewed.

Suppressing Individuality

Industrialism increases general insecurity in several ways. As already noted, increases in international trade make the availability of raw materials, safe trade routes, and open markets matters of security for people in industrial societies. Many aspects of international security, such as oil embargoes, foreign subsidies, product "dumping," and protectionism, hit home quickly.

Industrialism causes general insecurity through the production of pollution that threatens people's health. This includes smog, nuclear radiation, and chemical wastes. People worry that increased cancer rates may be pollution-related.

Fear of violence causes much insecurity. Industry mass produces devices designed to destroy things violently. These include rifles, shotguns, hand guns, dynamite, assault rifles, hand grenades, plastic explosives, and many more. Mass production makes many such devices inexpensive. Technological improvements increase their power and reduce their size and weight, making it easier to carry and conceal them.

Insecurity in the face of violence is increased by industrial interdependencies. Bombing an electric power station can leave millions of people without electricity, affecting domestic life, health care, and jobs. The explosion at New York's World Trade Center in February 1993 interrupted work for tens of thousands of people and threatened to cost hundreds of millions of dollars in lost business. It closed commodity exchanges affecting millions of people around the world.

In sum, industrialism increases insecurity partly because techniques of violence become more effective and available while the impact of violence is more widespread.

People react to insecurity by attempting to control its source. In a society dedicated to technological improvements and commercial interdependencies, people do not concentrate on controlling technological innovations or reducing interdependence. Instead, they concentrate on *controlling people* who may be tempted to put available technologies to disruptive use. Just as industrialism requires standard weights, measures, currencies, and radio frequencies, it requires standard people.

As industrialism spreads so does state-supported public education, which inculcates standard values, standard scientific views of reality, standard interpretations of history, standard hopes, and standard skills. People thus equipped "fit in," like replacement ribbons in typewriters. Others are misfits.

The limits of tolerance for misfits decrease with industrialism. Increasing percentages of people are institutionalized, mostly in criminal and mental institutions.

Feeling insecure, and sensing their loss of autonomy, many people in the United States assert the right to keep and carry firearms. This decreases general security and justifies additional government efforts to standardize people.

In sum, suppressing people accompanies industrialism's suppression of the earth. Victory over unruly people is called for by victory over unruly nature. We should for this reason question the received opinion that we have more freedom than people in technologically simple societies. Many anthropologists claim that individuality is greater in indigenous than in industrial societies. That is a topic for later.

Rather than eat standardized food at McDonald's, Patricia and I went to a hotel restaurant where we could get (what we were told was) traditional Bavarian food. We met an elderly couple who approached us because we were speaking English. He was Belgian and spoke of his experiences during World War II fighting with U.S. troops at the Battle of the Bulge. His wife was German and was among those whom German troops were protecting from invasion. The man ordered for us the best beer I have ever tasted, and we did not miss McDonald's.

Chapter 4

Nationalism, Bureaucracy, and the Holocaust

The Importance of Government

Our stay in Bavaria was most enjoyable. After visiting Garmisch-Partenkirchen, we spent some time near Füssen, an attractive market and tourist town. In and near the town are castles built by nobles of different centuries. The most famous is the most recent, Neuschwanstein, built under the direction of King Ludwig II of Bavaria in the 1870s and 1880s. Although incomplete at the time of Ludwig's death in 1886, its exterior was used by Walt Disney as the model for the castles in *Sleeping Beauty* and at the Disney theme parks.

Two-hour waits are common in the summer for those wanting to visit Neuschwanstein. I am not sure why castles fascinate so many people. Perhaps many people, craving larger dwellings for themselves and more luxuries, are voyeurs of wealth. Or castles may attract people because they are physical symbols of government power, which is increasingly important in our lives.

As industrialism makes the world more dangerous, governments are called upon increasingly to protect us. Industrialism fosters the acquisition of more physical possessions by more people. These possessions are protected by police, courts, and expanding penal institutions.

Governments protect people also from dangers that industrially produced possessions may pose. Automobiles and guns, for example, are inherently dangerous because they can so easily be used to harm people. Some dangerous children's toys are banned altogether. Other products, such as meat, medications, and airplanes, must meet government safety standards.

Many services, too, must be government-certified as safe. Few people in industrial societies can protect themselves from incompetent specialists, so pilots, physicians, electricians, and many others must be licensed by the government.

Industrial manufacture can be dangerous to workers and the public because it often employs substances that are health hazards, such as asbestos and DDT. In the United States, the Occupational Health and Safety Administration (OSHA) and the Environmental Protection Agency (EPA) are among those who address these dangers.

Governments are called upon increasingly to assure employment, because in industrial societies few people can assure employment for themselves. Considerable investment is often required for job creation. Governments are expected to undertake such investment, or encourage private investment, as with tax breaks.

Because many industries operate worldwide, jobs are protected from "unfair" foreign competition through international agreements, and by import taxes on goods "unfairly" subsidized by foreign governments. International agreements also encourage international trade to spur economic activity that creates jobs.

When people are jobless nevertheless, governments provide financial support for the jobless, or the government becomes their employer. With so many tasks to perform, governments are major employers in any case.

Unemployment is a concern partly because the interdependencies of industry and commerce make society vulnerable to physical attack, while industrially manufactured items make attacks easier. Providing employment is one way that governments forestall violence by those who are desperately poor.

Government programs of education and training prepare people for work. Standard educational backgrounds foster common values and outlooks that help people understand and get along with one another. Like programs that supply employment, education forestalls antisocial behavior. Governments (attempt to) forestall antisocial behavior also by (attempting to) limit access to drugs that are believed to interfere with social cooperation.

When people are too old to work, they receive government pensions. Family structures and spending patterns characteristic of industrial societies do not allow many old people to live entirely on the income of younger relatives.

These are only some of the tasks currently undertaken by governments. For the most part, governments are larger now than ever before due to industrialism. If free markets are beneficial because they foster the expansion and intensification of industry, as proponents maintain, it is ironic that free marketeers should decry government growth, which is an effect of increased industry and often spurs further industrialization.

Unfortunately, concentrating power in government leads, in our cultural context, at least, to increased human oppression. Paradoxically, attempts to master nature in the human interest tend to harm people more than help them. Commerce and industry, which are supposed to serve humanity by reducing scarcity, create insecurities that justify not only increased concentration of power in nation-states, but the use of that power against people.

The Importance of Nationalism

As government becomes more important, the struggle to control it intensifies. Representatives of different industries, seeking government subsidies, tax breaks, or trade concessions, vie with one another for attention. Labor, management, and consumer groups attempt to influence laws and regulations.

People compete for government favors because they realize that *economic and social positions are largely controlled by a small group of human beings.* This is contrary to the free market ideal. Adam Smith and his followers maintained that governments should support a market where, through immutable laws of supply and demand, economic results reflect the cumulative effect of myriad independent human decisions. No individual or small group would, under these conditions, determine results.

This conception of market activity had the advantage of deflecting criticism from wealthy people. Their social advantages were the result of impersonal forces that no one could control.

Increasingly, however, people realize that impersonal forces do not control the economy. Deliberate taxing decisions affect industry prices. The power to tax is the power to destroy. Government subsidies tend to lower prices as taxes raise them. Government safety requirements raise the cost of production.

The price of labor, too, is an artifact of government decision. Labor costs tend to rise as governments grant generous unemployment and welfare benefits that enable people to stay out of work if wages are too low. Legalizing unions, strikes, and secondary boycotts also strengthens the hand of labor and raises labor's price. Allowing companies to fire strikers and hire permanent replacement workers strengthens the hand of management and lowers labor's price.

Philosopher and sociologist Ernest Gellner puts the point well:

> Certainly the notion of a just price, inscribed into the nature of things, is a superstition. . . . But, alas, there is no market price either. . . . Once it is fully clear that the market operates only in a political context, . . . it also becomes obvious that the verdict of the market, the price, is not issued by the oracle of "the Market" alone. In fact it can only be the ventriloquist's mouthpiece for the particular political situation which happens to underlie it.

So divisive competition for government favors is common.

Social division must not be allowed to go too far, however, because industry requires cooperation before any product is complete, and dissidents can be particulary disruptive in industrial societies. Public education and other forms of mass communication foster social solidarity, often in the form of patriotism or nationalism, so that when times are hard people do not turn disruptively against one another or the government. The general message is that some others, outside the country, are responsible for problems. Arabs interfere with oil supplies, communist insurgents threaten free trade, and foreign government subsidies ruin fair competition.

It is a short step from blaming outsiders to blaming people within society who resemble outsiders in race, ethnicity, national origin, or religion. There is pressure on governments to expel "outsiders," restrict immigration, and attend exclusively to the needs of "real" Americans (or Germans, or French, or British, as the case may be). In the last decade, immigration and/or asylum laws have stiffened in all of these countries.

Corresponding to this, ethnic groups want their own countries. Because government favors are today's manna and they are disbursed with ethnic prejudice, the best way to protect oneself and one's group is to have one's own government doing the dispursement.

Another reason for ethnocentrism may be psychological. Many people feel insecure due to international dependencies and powerful multinational corporations. They crave the comforts of small-group solidarity and familiarity, but are surrounded by standard products, from soft drinks to computers to musical entertainment, and standard procedures, from manufacturing to accounting. Ethnocentrism may foster comfort (and solace when times are particularly bad) in a world that seems foreign and hostile.

Ethnocentrism can then be self-reinforcing. As one group asserts its right to control its own destiny, it often puts members of other groups in jeopardy, for example, by denying them equal employment opportunity. Members of the second group, made increasingly insecure, are more motivated than ever to assert their own right to ethnic insularity.

King Ludwig's castle symbolizes these tendencies. Ludwig lived when Bavaria was losing autonomy to the newly emerging German state, dominated by Prussia. The new state was promoted in part by appeal to sentiments of pan-German nationalism, which Ludwig shared. He was a patron of Richard Wagner due to the composer's interest in Aryan legends and distaste for foreign influences. As Patricia and I viewed Ludwig's castle from outside, demonstrations and violence against "foreigners" were erupting again in the northeastern German city of Rostock. It seems that a result of industrialism is often humanity turned against itself.

As we left Füssen for Munich on our last day in Bavaria, we did not realize how close our route took us to the prison where young Adolph Hitler composed *Mein Kampf*.

Dachau and Anti-Semitism

The highway to Munich airport, where we were headed to return our rental car, passes Dachau, a suburb of Munich. We decided to visit the museum that is now on the site of the infamous concentration camp.

Except for the barbed wire on top, the white walls around the camp could have contained a monastery or boarding school. I suppose in some areas of the world today monasteries and boarding schools have barbed wire, too.

The centerpiece of the museum is a large building at one end of the compound that used to serve administrators and visitors. Internal walls were removed to create one vast room dominated by blown-up photographs of documents and events. These depict the Holocaust from early manifestations of anti-Semitism, through conditions of incarceration and methods of extermination, to evidence found by the allies at liberation. Some artifacts, including means of torture, are displayed as well.

Some other buildings are models of barracks where inmates were housed. The wooden structures that served as beds suggest crowding. Most of these barracks are today represented only by outlines of their foundations on the ground. There may have been thirty or forty; I did not count.

There is a beautiful memorial fountain between the main museum building and the barracks. Other memorials are at the far end of the camp behind the housing units. In the corner is a passage through hedges to the crematorium. It was solid-looking red brick that resembled from the outside an old-fashioned triple-car garage.

I did not know until this visit why Dachau is so special. It was the first concentration camp and was used early in the 1930s, long before there were plans to exterminate Jews, to house communists, socialists, dissident clergy, gypsies, gays, and vocal opponents of national socialism. But the concentration camp is known today primarily for its role in the Holocaust, the deliberate killing of six million Jews under Nazi control.

Why were Jews singled out? I heard this question often while growing up Jewish in a New York suburb. It seems that several major elements of Western industrial culture were involved.

One consideration was religion. Until the second half of the twentieth century, Jews were blamed in official Catholic teaching for the death of Jesus. Furthermore, their continuing existence implied continuing rejection of Jesus, a sign of moral corruption. For these reasons, anti-Semitism persisted in many parts of Europe among people who had never seen a Jew.

Jewish roles in government also played a part. Because Jews were social outcasts, they often needed the special protection of kings, who sometimes required for this protection that Jews perform unpopular state functions, such as collecting taxes. This further tarnished the popular image of Jews.

Disruptions of traditional life caused by commercialism sparked anti-Semitism, too. When old securities disappear and outcomes are disappointing, it is satisfying to blame people already believed morally corrupt. Blame was placed on Jews with special ease because they were forbidden to own land in Central Europe before 1848 and were barred from many trades. Finance and business were among their relatively few opportunities, so they were prominent in the commercialism that disrupted traditional life. The self-fulfilling assumption of Jewish immorality came full circle in taunts that Jews were no good at growing or making things, only at buying and selling the work of others. In Poland the upper class disdained industrialization, so Jews were prominent in industry. This encouraged Poles to blame Jews for the evils of industrialization.

The success of Jewish merchants was also the envy of many Christian competitors. Jewish merchants in Vienna, for example, ignored customary practices. They expanded store hours, lowered prices, and began offering "specials."

As we have seen, resurgent nationalism is one result of industrialization. Jews were seen as an international people without a geographic base. Their neighbors associated moral integrity with geographic roots.

The new nationalism incorporated ideas from modern science to support racism. Linnaeus, the father of scientific taxonomy, assumed that just as species behave differently due to underlying biological differences, so do different races. His "biological" portrait of black Africans was not flattering. Central and Eastern Europeans concluded that Jews are inclined racially to moral inferiority. The theory of evolution and modern methods of measuring the skull were used to "prove" that Jews are inferior because they are less evolved, more apelike, than Aryans.

Modern science proceeds from and reinforces a zeal for improvement. Weeds should be eliminated by farmers and malignant cancers by surgeons. People who are cancers on the nation should be removed as well. Paul Joseph Goebbels, Hitler's propaganda chief, argued, "The fact that the Jew still lives among us is no proof that he also belongs with us, just as a flea does not become a domestic animal because it lives in the house."

This reasoning makes sense only in a society that expects "progress" and promotes it through scientifically engineered programs of improvement. In societies where the human condition is accepted as found, the mass extermination of "household pests," such as fleas, has no place. So the analogous extermination of people makes no sense.

In several ways, then, the impulse for the Holocaust depended on combining major elements of Western culture: a particular interpretation of Christianity that fostered anti-Semitism; the disruptions in social life caused by commercialism and industrialization; nationalism as one response to these disruptions; science that supported racism; and the belief that human life can be improved through social application of scientific principles. We have seen that these ideas, movements, and practices often justified human oppression. The Holocaust is just an extreme example.

A holocaust requires more than the Western ideas, movements, and practices just mentioned. Unfortunately, the missing elements are also part of Western culture. They are the distinctively Western separation of facts from values, almost exclusive reliance on means-end reasoning, and personal distancing fostered by modern technologies and divisions of labor. These come together in the operation of government bureaucracies.

I now explain why the Holocaust required a bureaucracy, what bureaucracies are, and how bureaucracies facilitate extreme inhumanity. Then a final assessment can be made of the relationship between Western culture and the Holocaust.

The Inadequacy of Hate

Hating a group of people and desiring their elimination does not lead to mass extermination without bureaucratic organization. Nazis discovered in 1933 that despite widespread anti-Semitism there was little popular enthusiasm for a planned anti-Jewish boycott, so they limited the action to one day.

More than five years later, after considerable anti-Semitic propaganda, Nazis were again disappointed by popular reactions to *kristallnacht,* an organized, widespread pogrom that got its name from the sound of breaking glass that accompanied nighttime attacks on Jewish homes and businesses. About one hundred Jews were killed. Unfortunately for the Nazis, many Germans showed sympathy for Jewish neighbors attacked by thugs. Himmler complained as late as October 1943 that good party members who agreed that Jews should be exterminated made exceptions for the Jews of their personal acquaintance. "Each one has his decent Jew. Of course the others are swine, but this one is a first-class Jew."

These reactions and sentiments suggest that Hobbes was wrong. People are not naturally selfish and indifferent to the suffering of others. Instead, people tend to care about and take responsibility for those in their community. According to one observer in Beirut, reactions in 1983 to the collapse of the Lebanese government illustrated this well.

> Beirutis . . . seemed to disprove Hobbes's prediction that life in the "state of nature" would be "solitary". . . . Rather, the behavior of Beirutis suggested that man's natural state is as a social animal who will do everything he can to seek out and create community and structures when the larger government or society disappears. Beirut was divided into a mosaic of neighborhoods, each tied together by interlocking bonds of family, friendship, and religion. . . . These microsocieties . . . helped to keep people alive, upright, and honest. . . .

Like the people of Beirut, German anti-Semites, even Nazi Party members and S.S. officials, were essentially social and moral in relation to people who were proximate to them. When Jews were proximate, they were included, too. This is why, despite widespread anti-Semitism and years of government anti-Semitic propaganda, Jews could not be eliminated through violence that reflected personal animosity. The Holocaust depended on conditions maintained in modern bureaucracies.

The Nature of Bureaucracy

Division of labor is essential in modern life. Specialization in production promotes industrial efficiency and creates needs in each person for goods produced by others, thus stimulating commerce. Specialization also fosters expertise in the service of scientific, technological, and artistic progress. "Bureaucratism" is specialization and division of labor in government.

Turn-of-the-century sociologist Max Weber described ideal bureaucracies as including "Precision, speed, unambiguity, knowledge of the files, continuity, discretion, unity, strict subordination, reduction of friction and of material and personal costs. . . ." These characteristics facilitate "the optimum possibility for carrying through the principle of specializing administrative functions according to purely objective considerations . . . ," allowing for "a discharge of business according to *calculable rules* and 'without regard for persons.'"

Bureaucracies of this sort reflect several features of modern Western society. "Unity" is needed so that a single product emerges from the efforts of different specialized people. This unity is achieved through "strict subordination." Efficiency ("precision, speed, and reduction of . . . costs") in the service of human good justifies subordinating some people to others.

Conducting business according to "calculable rules" reflects the modern scientific view that reality can be described mathematically. Calculable rules yield decisions on "purely objective considerations" because objective reality is amenable to calculation. Exclusive attention to objective considerations requires ignoring everything subjective. This includes whatever is idiosyncratic about individual people, hence the importance of operating "without regard for persons". Morality, too, must be ignored, as modern science and philosophy consider it subjective.

The Importance of Bureaucracy

These characteristics of bureaucracy were essential to the Holocaust. First, bureaucratic decisions are means-end oriented. There are no moral evaluations of the means (values are not part of objective reality), so changed conditions can lead to the substitution of barbaric for relatively benign means. This helped produce the Holocaust.

Following racist beliefs that claimed support from modern science and adopting the modern view that people can manipulate reality for human good, Hitler set a government goal. This was to make Germany *judenfrei*, that is, clear of Jews. Hitler did not specify the means to be used toward this end.

The first method was to deport Jews to neighboring countries, but neighbors were not willing to take Jewish refugees and the number of Jews in the Reich increased as the country expanded through conquest. The next idea was to create a separate Jewish principality in central Poland, but Nazi authorities in charge of administering conquered Polish territories complained that they already had enough Jews to police.

As conquest expanded further, the goal shifted from a *judenfrei* Reich to a *judenfrei* Europe. With France defeated, it was thought her colony of Madagascar could serve as a haven for Jews. This project failed due to the British Navy's control of the high seas. Only then did Nazis turn to the option of exterminating Jews as a means of getting rid of them. Rational means-end reasoning made this the logical next step. When morality is not part of rationality, extermination may be rational.

Having decided on extermination, bureaucrats in charge of the project consulted experts to determine the appropriate means. At first victims were killed by machine guns at point-blank range, but this proved demoralizing to soldiers. Proximity breeds moral attitudes. Even when the distance was increased to the maximum compatible with accurate aim, soldiers were upset. Taking this human factor into account experts came up with mobile, and finally stationary, gas chambers. The actual killers could then more easily see themselves as ordinary pest-control engineers. The bureaucracy used modern technology to overcome an obstacle to extermination by substituting distance for proximity.

Modern bureaucracies are adept at making conscience-numbing, reassuring separations. Jews were first separated conceptually by means of definition. This served the bureaucratic imperative of precision, and reassured non-Jewish Germans that they were not in jeopardy, thereby fostering popular indifference. Defining the target population also helped clarify whose job it was to deal with "the problem." Others were warned not to interfere.

Separation then took the form of physically transporting or enclosing Jews to minimize proximity between Jews and Germans. Stanley Milgram showed in his experiments that most people will in the name of science follow authoritative orders to the point of inflicting pain, and even death, on an unknown, unseen "research subject." But compliance is reduced quickly as physical distance from this "subject" is reduced. Anticipating Milgram's findings, Nazis found that physically separating Jews facilitated general indifference.

Of course, those in the bureaucracy of extermination could not be completely detached because they dealt with Jews. But here, too, a kind of separation was used that had the same conscience-numbing effect. Most of those who organized the extermination spoke on the telephone, wrote memoranda, and engaged in other actions typical of bureaucratic work in general.

Others, who were more involved physically, had their conscience soothed by the division of labor that characterizes modern organizations. Most workers engaged in benign activities, such as fixing train tracks, hauling fuel, laying bricks, and the like. They did not typically see the ultimate results in gas chambers or crematoria. They were like most factory workers making napalm who rarely take personal responsibility for children burning to death.

Where morality does enter, it is bureaucratic morality, which consists primarily in doing one's job well. Due to the division of labor, workers in organizations count on one another to play their part lest the entire enterprise fail. Also, people take pride in abilities associated with their areas of expertise. Accordingly, those trying to design mobile gas vans were disturbed by criticisms and complaints.

There was no moral evaluation of the organization's goals. Any such evaluation would conflict with the scientific and philosophical belief that values are subjective, whereas organizations must be guided by objective criteria. Moral evaluations of the organization's goals would also jeopardize needed bureaucratic subordination. Thus, according to sociologist Zigmunt Bauman, "The Nazi mass murder of the European Jewry was not only the technological achievement of an industrial society, but also the organizational achievement of a bureaucratic society."

Moral Progress

Belief in moral progress is central to much Western religion. Things may be going badly now, we are told, but eventually the forces of good will defeat the forces of evil. The messiah will come, and the Battle of Armageddon will be won, and the world will be perfected. This vision is secularized when actions decisive for improvement are put in human hands. Medical science will cure disease and governments will control crime.

Both religious and secular versions expect civilization to promote progress. Yet the Holocaust depended on ideas, movements, and institutions characteristic of modern, Western civilization. The ideas include the beliefs that progress can exist on earth, that human beings can engineer such progress, that nation-states should organize the engineers, that means and ends can be evaluated separately, and that facts and values are separate. Influential movements included those toward technological mastery of nature, producing powerful means of destruction, division of labor and specialization of expertise, and commercial and industrial interdependence, creating economic dependence and insecurity. Important institutions include sovereign states and bureaucratic organizations.

Together these made the Holocaust possible. They were necessary, but not sufficient, conditions. More generally, they facilitate immorality, as more recent events suggest.

Serbs, who were implacable enemies of German Nazis, practiced ethnic cleansing in the early 1990s to create an ethnically uniform state. Some people maintain that Serbian aggression stemmed from ancient animosities that no one could control. However, Tom Sjelton, who covered the war in Bosnia for National Public Radio, found that the violence resulted from an orchestrated campaign by Serbs seeking political power. A Human Rights Watch report claims that political manipulation through mass media is responsible for recent Hutu violence against Tutsis in Rwanda as well. Political leaders resuscitate and play upon longstanding prejudice to garner popular support. Far from being uncontrollable, the violence results from political manipulation and bespeaks the current importance of political power.

The importance of nation-states is reflected in the idea that national sovereignty should be respected. States should not interfere in one another's internal affairs. This permitted Saddam Hussein of Iraq to gas Kurds who wanted a state of their own.

Perhaps most ironically, Jews, who understandably wanted a state of their own for self-protection, established a state that favors Jews over Arabs. Arabs are restricted more than Jews in land ownership and use. Illegal land use is often enforced selectively against Arabs but not Jews. There is no holocaust in Israel, but there are, or have been, holocausts in Iraq, Bosnia, Nigeria, Ethiopia, and Cambodia, to name but a few. Although not all of these countries are industrialized, the influence of industry and commerce, supplying weapons and insecurity, can be seen in all of these tragedies.

In spite of these examples of moral failure, belief in moral progress is so ingrained in our society that even some critics of industrial, human-centered values accept it. Aldo Leopold, who is often considered the father of modern ecology, endorsed nonanthropocentric, nonindustrial, and noncommercial values in "The Land Ethic." Yet he begins the essay, published after the Holocaust, proclaiming moral progress from ancient Greece to the present.

J. Baird Callicott, a contemporary expositor of Leopold's thought and champion of the land ethic, defends claims of moral progress. He acknowledges that moral practice has not improved, but claims that "moral consciousness is expanding more rapidly now than ever before. Civil rights, human rights, women's liberation, children's liberation, animal liberation, etc., all indicate . . ." accelerating rates of progress in moral consciousness. This reasoning resembles pie-in-the-sky religion. Because life on earth is hard and injustice abounds, some people who consider themselves traditional Christians attach importance to heaven instead of earth. Callicott attaches importance to "moral consciousness" to defend claims of progress against the evidence of "moral practice."

Why should we care about moral consciousness? Callicott mentions civil rights. However, regardless of income level, most blacks continue to live in highly segregated communities. Racial discrimination in the real estate and financial communities persists and accounts for much of this segregation.

Income is also a problem. In my town of Springfield, Illinois, for example, per capita income for whites is twice that of blacks. Given these realities, our society's self-congratulations for achievements in civil rights may be not only premature, but misguided. They may perpetuate a myth of progress and discourage attention to persistent problems.

Callicott mentions "children's liberation." A Carnegie Foundation report reveals that during 1993 child neglect and poverty increased significantly in the United States although the country as a whole was considered prosperous by conventional economic criteria. "Childhope estimates that the number of street children worldwide exceeds 100 million," notes American free-lance writer Germaine W. Shames. "In parts of Asia, perfectly healthy children are maimed and sent out to beg."

The same lamentable performance exists in another area mentioned by Callicott—"human rights." While slavery is condemned universally, an estimated two hundred million of the world's people are slaves at this time.

Optimists might believe that improvements in moral consciousness will eventually produce improvements in moral practice, but since history does not yet support this belief, it amounts to no more than typically Western blind faith in progress.

Progress in moral consciousness may serve, instead, as symbolic ballast, that is, symbolic compensation for inadequate performance, as in the phrase "whistlin' Dixie." This applies to a person who is defeated, as the South was in the Civil War, and who blunts the psychological impact of defeat by pursuing activities symbolic of pride and continued struggle. In general, symbolic ballast helps people maintain psychological equilibrium through symbolic gestures at variance with, and therefore in the face of, realities that are difficult to accept. For example, I know a generous man who is symbolically selfish, through adoption of Ayn Rand's philosophy, to relieve the stress of generosity.

Symbolic ballast may be related to Aristotle's doctrine of the golden mean. Aristotle maintained that the proper action is a mean between extremes. People inclined toward one extreme may best attain the mean by aiming at the opposite extreme, rather than at the mean. Selfish people may best cultivate proper generosity by aiming at more generosity than is really warranted.

I suggest the *possibility* that symbolic adoption of one extreme often serves to encourage actions at the opposite extreme. In the present context, ever more extreme statements of moral inclusion may combine symbiotically with ever more extreme actions of unjustified violence. Progress in moral consciousness may be co-dependent with continued moral degeneration, the way that "helpful" actions of the "good" spouse often foster continuation of addictive behavior.

I offer the theory of symbolic ballast to account for simultaneous "progress" in moral consciousness and barbarity in moral practice. It seems as plausible as the claim that progress in moral consciousness leads to progress in moral practice.

Viewing our species' lamentable history as any kind of moral progress bespeaks (typically Western) blind faith in progress, and substituting moral consciousness for moral practice as the philosopher's primary concern bespeaks Western religion's replacement of heaven for earth as the venue of importance. Unsubstantiated claims of progress by leading environmental philosophers shows that some potentially dangerous Western thinking is so ingrained that it even affects the counterculture.

Departure

We left Dachau puzzled, disgusted, and fearful that it could happen again, to Jews or to others. A month earlier, eight hundred young Nazis attacked a refugee housing complex in Rostock, a northern German city. Police did little to stop the mob or protect the refugees. German government officials later joined demonstrations against violence and intolerance, but soon acceded to key Nazi demands, agreeing to remove all refugees from Rostock and to deport one hundred thousand Gypsies to Romania. Other foreigners would be removed to Poland, Czechoslovakia, and Bulgaria.

Jewish cemeteries were desecrated with swastikas, a concentration camp memorial was burned, and Danish Boy Scouts were beaten. Official German statistics for 1992 attribute to neo-Nazis forty-five hundred attacks on foreigners, resulting in hundreds of injuries and at least seventeen deaths. One of their "intellectual" leaders maintains that "The Holocaust is a fabrication, the pictures of the dead, of gas chambers, of mass murder are filmed by Hollywood, narrated by Trevor Roper, and directed by Hitchcock." Has this person ever been to Dachau? He lives twenty minutes away in Munich, where we caught a train for Austria.

Chapter 5

Nuclear Power
and Radiation Exposure

The Hearing Begins

Our train arrived in Salzburg on Sunday about 5:30 P.M. I was concerned that we would not reach our room before 6:00 P.M., and was unsure if that jeopardized our reservation.

I had written ahead to the organizers of the World Uranium Hearing to request a decent bed and breakfast that was close to the hearing. They replied with the name, address, and telephone number of Peter Jecel (pronounced "Yetzel" I later learned). I telephoned Mr. Jecel to say that we would arrive after 6 P.M.

I reached an answering machine and was stating my message when Mr. Jecel came on the line to tell me he would pick me up. This was surprising; I had never heard of a bed and breakfast with shuttle service.

It turned out that the Jecels do not have a bed and breakfast. They are two physicians who volunteered to provide a free place for the week for people attending the hearing. Patricia and I were the lucky recipients of their generosity.

Nearing fifty, the Jecels were just a year or two older than us. They have two daughters, Julia and Maria, nearly the same age as our two older daughters. We had the pleasure of meeting them at week's end when they came to visit from Vienna where Julia studies medicine and Maria art.

The whole family exuded warmth and rootedness. Both Christa and Peter Jecel are native Austrians, and Christa's mother lives in Salzburg. Their sense of rootedness reminded me of what I was told by a colleague in anthropology. Most Europeans are indigenous to Europe. Homo erectus entered Europe 750,000 years ago, and "people" have lived there ever since. I knew that we were going to Austria to hear indigenous people speak about nuclear issues, but it had not occurred to me that the first would be Austrians.

Austria was chosen for the World Uranium Hearing in part because it is nuclear free. Austrians do not have, and have never had, either nuclear weapons or a working nuclear power plant. The one nuclear power plant they built was rejected by popular referendum before it could generate electricity.

Another reason for choosing Austria is the presence in Vienna of the headquarters of the International Atomic Energy Agency (IAEA). Amid increasing concern about the safety of nuclear energy, the major international regulatory agency is located where officials and their families are least exposed.

Salzburg was an excellent location because it is close to Munich, where the hearing was organized, and because the city was generous and receptive. The Residence Palace, where the archbishops who ruled the province of Salzburg lived, was lent to the hearing free of charge, and the local museum had a moving display related to uranium issues.

The city seemed to me especially appropriate because its transportation was organized to save energy. All energy savings undercut reasons for nuclear power. There were both bicycle and walking paths, and an efficient bus system, that minimize automobile use. Patricia and I sometimes took the bus between the Jecel's and the university in the old city where the hearing was held. But we also walked, as it was only about two kilometers.

Testimony began Monday morning. Peter took off the week so

that he could attend. He and I took bicycles to a stately building where people were crowded around a long table on the ground floor to get their official name tags and updated schedules. On the second floor was a spacious area for coffee and informal chats outside the auditorium where testimony was given.

Entering the auditorium, we picked up earphones and received instructions on language translation. The atmosphere was what I had always imagined at the United Nations. People finding seats were clearly of different racial and cultural backgrounds; many dressed in non-Western styles.

We were to hear about the nuclear generation of electricity, which promised originally to provide people with virtually unlimited, inexpensive physical power to manipulate nature in the human interest. In the 1950s an official of the U.S. Atomic Energy Commission said this power would be "too cheap to meter."

Unfortunately, enormous human oppression accompanies this ambitious attempt to control nature. The present chapter begins with uranium mining, the first step in nuclear power generation. Many of those oppressed are indigenous people, but as this chapter and the next indicate, almost everyone is jeopardized by activities associated with generating electricity by nuclear fission.

Dangers of Radiation

Concerns about safety have accompanied nuclear power from the start. Private insurance companies refused to fully insure nuclear power plants. The near disaster at Three Mile Island in 1979 alerted many people to dangers that originally worried insurance companies, and the Chernobyl disaster confirmed some of the worst fears. But the death-dealing nature of nuclear power would exist even if power plants were perfectly safe, because mining uranium exposes people to dangerous radiation.

The detrimental health effects of radiation on human beings has been recognized only gradually. When I was young, shoe stores had x-ray machines that enabled people to look through the top of shoes they were wearing to check the fit. Pregnant women used to be given x-rays, much as many now have sonograms, to monitor the progress of gestation. These practices have disappeared because exposure to radiation is now considered too dangerous.

Dr. Alice Stewart, a British physician who appeared to be in her late 60s, was among the first speakers at the hearing. Receiving the kind of respect, almost reverence, that a priest might receive in a traditional society, she outlined her pioneering research that helped prove radiation's dangers. She noted that ill-health effects of radiation, such as cancer, are difficult to identify when radiation doses are low. There is a long delay between exposure and illness, and the illnesses involved occur in our species anyway. Nevertheless, Dr. Stewart provided excellent evidence that small amounts of radiation stemming from twentieth century technologies can be harmful to human health. She compared three groups of people: 1) children who died of cancer before age ten, 2) children who died of other causes before age ten, and 3) children who survived to age ten. She found that cancer victims differed from others in just one statistically significant way—their mothers had, on average, been exposed to twice the number of abdominal x-rays during pregnancy.

Due to this research and others of its kind, a U.S. National Research Council (NRC) committee concluded in 1989 that the cancer-causing potential of acute radiation is three or four times greater than had been thought only a decade earlier. The International Commission on Radiological Protection (ICRP) has progressively lowered permissible levels of radiation exposure in the workplace. Its level in 1990 was only one-fifteenth of what it had been in 1934.

Recent evidence favoring additional lowering comes from a study of nuclear workers at Sellafield, England. Children of male workers were six to eight times more likely than others to contract leukemia, even though the fathers' exposure was within current safety limits.

Uranium Miners

Because people are so sensitive to radiation, uranium miners are at significant risk. According to Dr. Gordon Edwards's testimony on Monday morning, the danger to miners was predictable without any of the epidemiological studies of the past fifty years. People knew centuries ago that miners working in the Erz Mountains in what is now the Czech Republic often died from an unknown lung disease. In the nineteenth century the disease was identified as lung cancer, and in the twentieth, uranium's radiation was recognized as the cause. In 1942, when uranium was sought for the war effort, cancers among miners were predictable. For national security reasons, this aspect of nuclear technology was covered up, resulting in a continuing belief, even among many in the industry, that nuclear power is safe.

Only gradually has the danger of radiation exposure during mining been officially acknowledged. In the mid-1970s Czechoslovakian uranium miners were found to have ten times the normal cancer rate. A study of French miners in 1983 found that they died of lung cancer 91 percent more often than those in other professions. This was unwelcome news in some French circles because France derives 75 percent of its electricity from nuclear power, more than any other country. Further investigation was undertaken, yielding in 1985 the worse conclusion that miners die of lung cancer 177 percent more than average. Symptomatic of the nuclear industry's financial and political power, this information was kept secret for three years.

India has the largest and most independent nuclear program in the Third World. On Tuesday evening at the hearing Xavier S. Dias, an anti-nuclear activist from India, testified that two doctors who treated uranium miners, "said that they were unaware of any radioactivity cases." It quickly became apparent why they were unaware. All miners' "blood samples [and] biopsies . . . are . . . sent to the Bhabha Atomic Research Center in Bombay, and those results of the blood tests on patients are never shown to the concerned doctors nor to the patients. It all goes under the Official Secrecy Act."

Given uranium mining's dangers, it is no surprise that poor, indigenous people are disproportionately represented among uranium miners. The Indian mines just referred to, for example, are worked primarily by the Santal, indigenous people in India's Jharkand region. China similarly does most of its uranium mining with labor from Tibet, which it occupied in 1949.

Indigenous people in the United States, too, are victims of uranium mining. Alveno Waconda is a Laguna Pueblo who testified at the hearing that he and others of his tribe worked at the Anaconda Mine in the southwestern United States. He recalled:

> The uranium ore was all around us, whether the wind blew it on us, from dusty conditions. There were times when we were having our lunch sitting on the highgrade ore stock pile. We had lunch sitting in loader buckets to get out of the hot sun. No one warned us that the buckets were contaminated from the ore.

As for safety inspections, he said, "somehow we always knew when the inspectors were coming . . . , giving us enough time to straighten up our area. . . . When we had safety meetings we were never given information on the dangers of radiation, there was only concern of equipment hazards." About 93 percent of these workers were Pueblo.

Native American Philipp Harrison told the hearing about the Red Valley mining facility where hundreds of Navajo men worked. "Early mines were very unsafe and dirty. There was no ventilation of these mines, no safety equipment, no respirators, no gloves were provided."

Similar conditions existed, according to federal mine inspectors, at the Shiprock mine owned and operated by Kerr-McGee on Navajo territory. One hundred fifty Navajo miners worked there during its eighteen years of operation. Within several years of its closing thirty-eight of those miners had died of radiation-induced cancers, and another ninety-five had serious respiratory ailments and cancers.

Uranium Mining as a Radiation Pump

Pollution of the area near the mine, including water resources, threatens the health of entire communities. Ninety-nine percent of mined uranium ore is left at the mine site, because only the richest 1 percent, in which 15 percent of radioactivity is concentrated, is worth hauling away. After on-site milling separates and concentrates the richest ore, 85 percent of unearthed radioactivity is left behind in tailings on the surface of the earth. Since milling pulverizes the rock to the consistency of flour, these tailings are subject to dispersal by wind. Power plant catastrophes aside, these tailings contain nuclear power's greatest addition of radiation to the human environment.

Radiation is dangerous, Dr. Gordon Edwards explained at the hearing, because it results from the spontaneous explosion of unstable elements, such as uranium. Although particles emitted in these explosions are called rays, they are really like shrapnel from a bomb. Radioactivity harms health when "the shrapnel rips through the cells of the body, and in the process breaks thousands of chemical bonds randomly. It's like throwing a grenade into a computer. The chance of getting an improvement . . . is very small, and similarly with radiation events and human cells."

Here are some facts about radioactive substances:

1) Each radioactive substance has a half-life, which is the time it takes half of the nuclei to decay. Different elements have different half-lives. Uranium-238 has a half-life of 4.5 billion years, whereas thorium-234's half-life is 24.1 days.

2) Because the process of radiation is a spontaneous one, it is not started and cannot readily be influenced (stopped, speeded up, slowed down) by human intervention.

3) Except in the case of polonium-210, when atoms of radioactive elements emit their rays, they do not cease to be radioactive,

the way bombs, once detonated, generally cease to be explosive. Instead, the atoms of one element are transformed into atoms of a different radioactive element. Uranium-238, for example, becomes thorium-234, which becomes protactinium-234, and so on, until lead-206, a stable, nonradioactive element, is finally reached. Each radioactive element has its own half-life, which may be longer or shorter than that of its predecessor.

4) The radioactivity of a "daughter" element can be several times that of its "parent." Within half an hour, for example, the decay of radon gas increases radioactivity fivefold.

Dr. Edwards concluded that uranium mining is "a mechanism for pumping radioactivity into the environment." This radioactivity lasts for millenia because "the effective half-life of the radioactivity [in mine tailings] is 80,000 years."

Physicist Peter Bossew, who also spoke at the hearing, translated radiation release figures associated with uranium mining into human deaths caused by this amount of radiation. In order to run a large, modern reactor for one year, an average of 440,000 tons of uranium rock must be mined. The vast majority of this, as we have seen, is left as tailings. Radon gas from these tailings will eventually cause an average of seventy-six radiation-induced deaths. Radium from the tailings seeps into groundwater, producing (over the years) another twenty fatalities. This is nearly one hundred deaths resulting from mining needed for the operation of each nuclear power plant each year. With more than four hundred such plants currently in operation worldwide, the mining needed for each year of nuclear energy use results eventually in forty-thousand human deaths from radiation exposure. Many of these will be in the United States and Canada, where nearly half of all uranium mining takes place.

Impact on Indigenous Communities

One reason the hazards of mining uranium are largely ignored is that in many countries mines are located near, and principally affect, poor, indigenous people. In the United States, 70 percent of uranium mining is done on Indian territory. The percentage is higher in Canada. As noted already, much uranium mining in India is done among the indigenous Santal people, and China does its uranium mining in Tibet. Australia principally uses aboriginal territory.

The negative effect on community health is sometimes obvious upon investigation. At the Shiprock mine run by Kerr-McGee on Navajo territory a radioactive tailings pile is just sixty feet from the San Juan River, the major source of drinking water at Shiprock and downstream communities. Birth defects in these communities were two to eight times the national average, and microcephaly is fifteen times what is normal. The same kind of water pollution and community health deterioration occurred at Tuba City, Arizona, also on the Navajo reservation, and at the Kerr-McGee mine at Churchrock.

Tsewang Norbu, who testified at the hearing on conditions in Tibet, said, "A high proportion of Tibetan villagers living near what is believed to be a Chinese uranium mine have died after drinking water polluted by waste from the mine. . . ." In Tibet's Amdo region fifty people died mysterious deaths in 1989 and 1990, and in the summer of 1990, "twelve women gave birth . . . and every child was dead at, or died during, birth."

Medical doctor Larissa Abrjutina, representing indigenous reindeer herders of the Chukchi Peninsula in Siberia, tells a similar story. "[I]n the Bilibino region . . . more that 70 percent of the people at working age suffer from oncological diseases."

Of course, communities of indigenous people are not alone at risk. Only once in forty years did the French government test milk samples drawn from cows grazed in the vicinity of uranium mines. Nevertheless, uranium mining, and all other forms of commercial-industrial development that pollute and despoil natural resources are, generally speaking, inherently more harmful to indigenous, than to other, people. The reason is that indigenous people are tied by their religion, mythology, traditional life-sustaining practices, and self-concept to a particular part of the earth. They are, to paraphrase Ms. Yazzie's poem that was quoted earlier, steadfast lovers of their particular land. If it is polluted or despoiled, they cannot move away and remain who they are. This contrasts sharply with people who lead an industrial way of life. The people who survived the pollution at Love Canal, a neighborhood near Buffalo, New York, where residential housing was built atop toxic industrial waste, could move away with their culture, language, religion, and self-concepts basically intact. The Navajo living near Shiprock mine in the southwestern United States, the Santal of India, the reindeer herders of Siberia, and other indigenous people cannot do the same. This is why many Navajos stay on ancestral lands even though reproductive organ cancer among their teenagers is seventeen times the national average.

Drawbacks of nuclear energy uncovered so far are: 1) Radiation harms people more than previously thought, 2) uranium mining, the essential first step in producing electricity through nuclear fission, unavoidably exposes people to huge amounts of radiation, and 3) indigenous people, who need and use little electricity, are often exposed most. Besides suffering illness, deformity, and early death, their entire way of life is ruined because their cultures depend upon residence in a particular area.

Creating Radioactivity

Operations after uranium is mined and milled aggravate matters. Yellowcake, a solid substance that results from milling, must be converted to gas in preparation for needed enrichment.

The conversion process can be dangerous. Lance Hughes of the Muskogee Nation in Oklahoma testified at the hearing about the conversion facility located near his people in Gore, Oklahoma. It is large, producing "20 percent of the world's supply for nuclear fuel raw production." Unfortunately, says Hughes, "because we are very culturally and geographically isolated," it has "been run very sloppily. . . ."

Mr. Hughes showed a videotape containing news reports related to contamination at, and from, the facility. On the tape John Montgomery, deputy regional administrator of the Nuclear Regulatory Commission (NRC), states that Sequoyah Fuels was for years dumping uranium on the ground underneath the building. At least 21,000 pounds of uncontained uranium was released within one mile of the Arkansas River. Other problems included a detector of workers' radiation contamination being turned off and unplugged.

In spite of all its findings, the NRC decided not to close the plant. The NRC official explained:

> The standard for shutting a plant down is, is there an imminent immediate threat to the public health and safety? And it's a very high standard. The central issues we were dealing with here were management issues and environmental issues. We did not believe that taken together they reached the standard.

This is how U.S. regulators protect public health from the dangers of handling uranium.

Because the dangers of uranium leaks and radiation exposures result in mostly illnesses and premature deaths that will not occur for ten or twenty years, the NRC's exclusive focus on "imminent immediate threat[s] to the public health and safety" virtually guarantees that most threats of handling uranium seldom will be adequately addressed. What is more, because the NRC is responsible for the safety of nuclear power plants in the United States, such plants, too, could pose long-term health hazards and remain open.

The Sequoyah Fuels operation in Oklahoma is just one facility in one country where the conversion process takes place. It is possible that other facilities are better managed, but it is unlikely that regulators in China and India, for example, where nuclear industries are shrouded in secrecy, are more responsive to public health concerns.

After conversion to a gaseous form, the uranium must be enriched. Most uranium contains only about 0.7 percent uranium-235, the rest being primarily uranium-238. But uranium-235, not 238, is fuel for fission, and a concentration of 3 percent is needed in most nuclear reactors. Enrichment helps to increase the percentage of uranium-235 to the level needed. I wondered why uranium-238, if it is radioactive, is not fuel for fission. Dr. Edwards explained that all radioactive atoms disintegrate through tiny explosions, but only some among them (uranium-235 but not uranium-238) are fissible. When bombarded with neutrons, uranium-235 "breaks apart into two large chunks, and in the process it gives off more neutrons," which can then break apart other atoms of uranium-235 to keep the fission process going. When the nucleus breaks apart, "it also gives off about 400 times as much energy as is given off by a radioactive event." In nuclear power plants, this energy, in the form of heat, is generally used much like burning coal to heat a liquid, often water, that turns a turbine that generates electricity.

People usually can control fission by controlling the bombardment of fissile material, such as uranium-235, by neutrons. We can start fission by starting neutron bombardment and can slow it down, or stop it, by interposing material that absorbs neutrons. Absorption prevents neutrons from hitting, and causing fission in, fissile material in the vicinity. Radiation, on the other hand, because it is a spontaneous process in certain materials, cannot easily be controlled or stopped. Due to this and its toxicity it is the major hazard associated with nuclear energy.

One of the worst aspects of nuclear fission is that the products of fission are highly radioactive. These products are entirely different from the elements involved in the series of spontaneous radioactive decay from uranium-238 to lead-206. The fission products include the large chunks that uranium-235 breaks into when bombarded with neutrons. Many of these fission products, such as strontium-90 and cesium-137, easily bond with organic molecules and enter the body, which they injure from within.

Another result of fission is the production of plutonium-239, which is basically uranium-238 with a neutron added through bombardment. Plutonium, however, is highly radioactive and has a half-life of 24,400 years. It is arguably the most toxic substance on earth.

The result of fission, then, is the dramatic increase in the radioactivity of the materials involved, the nuclear fuels being used in the reactor. Radioactivity results from atomic instability. Neutron bombardment and atomic breakups result in atoms of greater instability (radioactivity) than those that existed before. Radioactivity increases almost half a billion times, and after seven years of decay it is still almost a million times higher than in the initial uranium ore. Nuclear power, then, is produced by concentrating fissible materials that are radioactive and, through neutron bombardment, producing at the same time great energy releases and enormous increases in radioactivity.

This is a major reason why nuclear power plants can be so dangerous. If (when) nuclear material from the reactor core is released into the environment (as at Chernobyl in 1986), the contamination is unlike anything that would exist in nature without the intentional production of nuclear fission by human beings. Worldwide cancer deaths from Chernobyl are expected to number from 14,000 to 475,000.

International Conspiracy

People gave testimony at the hearing about cleanup efforts after the 1986 explosion at Chernobyl. Nikolaj Ostrogskij was sent to Chernobyl after the accident to head an airplane expedition that was trying to prevent precipitation in the Chernobyl area "to keep the radioactive dust from penetrating into the soil and reaching the ground waters and open waters." He reported that "for days the local population was not informed about the character of the accident, nor about the necessary protective measures to be taken." But "members of the Kiev administration secretly started evacuating their families," while "the laboratories in Kiev, where the instruments were kept which would have made a measuring of the degree of radiation possible, were closed down and sealed." These are only a few of the many examples he gave of immoral government duplicity after the accident. Another is this: "Five to six months after the explosion the Kiev administration decided to have the city cleaned of its contaminated autumn leaves. Kiev school children were sent to collect, and sometimes even ordered to burn, the radioactive leaves."

The most alarming testimony came from Vladimir Chernousenko, physicist and scientific coordinator of the cleanup, concerning reactions of the International Atomic Energy Agency (IAEA). International watchdog agencies are relied on to provide accurate information. According to Chernousenko, however, the IAEA joined the Soviet cover-up. The head of the IAEA came to Chernobyl and "saw very well that at that moment the fourth block was throwing out millions of Curies of activity," many times what the Soviet government claimed. But "at the same time he made calming statements: 'Just be quiet, everything is all right, everything is under control, there is no cause to worry.'"

Because of reactions like this, Chernousenko and his colleagues began to perceive their government and the IAEA as part of what he called "the atomic Mafia." Because they were all committed to the use of atomic energy to generate electricity, they had a common interest in minimizing dangers and reassuring the public. To do otherwise would risk public recognition that nuclear energy was a bad idea from the start. This would ruin reputations and job prospects for those making nuclear energy their life's work, would embarrass governments committed to nuclear energy, and could precipitate potentially disruptive political activity by concerned citizens. A conspiracy of silence existed among people and countries committed to nuclear power.

This conspiracy included the French government, which is heavily committed to nuclear power. I was in France at the time and noticed that television depictions of the spread of the nuclear cloud showed it stopping at the French border, as if confiscated by customs officials. While German and Swiss governments required that milk be discarded because the cows had eaten contaminated grass, just across the border in France the milk was considered completely safe. Safety in the nuclear age seems to be ruled more by politics than science or medicine.

This helps to explain the United States's NRC requirement that, before a facility is closed, there must be "an imminent immediate threat to public health and safety." A more prudent standard would require closing facilities associated with nuclear power, such as the conversion plant in Gore, Oklahoma. This would be not only expensive, but would erode confidence in the entire industry.

Direndra Sharma testified at the hearing on the heavy hand of politics in India. A film that her group had prepared, detailing duplicity and danger in India's nuclear policy and program, was censored by the government. Although the Indian Atomic Energy Commission claims never to have released radiation beyond permissible limits, it turns out that any release is permitted by the IAEA if it is "as low as reasonably achievable," where "reasonable" means "socially, economically, and technologically." As long as a country is doing its best with available money and technology, any radiation leak, no matter how large, is within IAEA limits, because it is "as low as reasonably achievable."

France has often risked a Chernobyl-type accident. In 1988, a nuclear plant at Flamanville lost its cooling capacity twice, and several others were shut down due to radiation leaks. In 1990 government regulators found blocked valves in the backup cooling system at two power plants, the Gravelines plant near the Belgian border and one at Dampierre near Dijon. Had backup cooling systems been needed, the result would have been a Chernobyl-type disaster. Subsequent investigation found similar problems at five other plants, yet French officials deny there is danger.

The Politics of Nuclear Waste

Political considerations are prominent in discussions and preliminary decisions about the disposal of high-level nuclear waste. As we have seen, human-induced fission processes create radioactivity levels that are many millions of times higher than what otherwise exists in nature. This is why high-level nuclear waste contains 95 percent of the radioactivity of all U.S. nuclear waste, although it is only 1 percent by volume. Ninety-five percent of this high-level waste is from civilian nuclear reactors designed to generate electricity. It is growing so rapidly that in 1991 it was three times what it had been in 1980, and twenty times what it was in 1970.

How will we deal with this waste? It seems that in the United States, Germany, and France, political considerations predominate over scientific judgments. In France the issue, like the waste itself, is currently too hot to handle. No sites have been named yet, and the current plan is to delay a final decision for fifteen years, while the high-level waste inventory nearly triples.

Ulrike Fink, a German biologist and member of Germany's Parliament, told the World Uranium Hearing that German authorities have chosen Gorleben, "a site which was given no chance by the experts before. . . ." It was chosen, she contends, "due to its location on the former German/German border." This enables former East and West German citizens to share equally the dangers of possible contamination from the high-level wastes. "It is curious: The more they investigate the more it becomes obvious that Gorleben is not suitable at all. Under normal circumstances, this would lead to giving up—but not in this case."

The situation in the United States is similar. Congress passed the Nuclear Waste Policy Act in 1982, requiring the study of several sites for disposal of high-level wastes and the eventual selection of two, one east and one west of the Mississippi. The eastern sites investigated all proved geologically unsuitable. In the west, Nevada was chosen, it seems, because its small population gave it weak representation in Congress. In 1986 Congress ordered the Department of Energy (DOE), which has jurisdiction over nuclear waste, to study just one site—at Yucca Mountain, Nevada.

When a government agency needs to certify a site as suitable, and only one site is being investigated, there is tremendous pressure to find that site suitable. Ulrike Fink explained that scientists investigating the Gorleben site for the German government assume the worst scenario possible—flooding of the depository, release of radiation, and so forth—and calculate resulting human radiation exposure. If this turns out to be above acceptable limits, "it does not mean that the site is not suitable, but the modelling and calculation has to be performed once more by using different input data or methods." Because researchers in the United States are under similar pressure to find a given site suitable, there is risk of similar intellectual dishonesty.

Such dishonesty can be disastrous. Consider the U.S. space shuttle *Challenger* that exploded in 1986 with seven crew members aboard. Bowing to political pressure to avoid additional launch delays, NASA and executives at a firm that designed critical components ignored warnings by responsible engineers. Afterwards, these engineers were fired, and years later were still unable to find employment in the aerospace industry. Can we expect the DOE to be more responsive to well-founded concerns when the possible disaster, even if it occurs, may be far in the future?

Unjust Distribution of Risks

In keeping with a pattern by now familiar, Yucca Mountain is part of the territory claimed by indigenous people, the Western Shoshone Indians. Their desires and those of other Nevadans are being overridden by Congress, which has removed from the state its right to oversee environmental permits. The creation of jeopardy has justified, yet again, concentration of power in a more central organization—the federal instead of state government—and subjection of local people to decisions of the more central power. Were the number of U.S. nuclear power plants to double in the next forty years, as proponents desire, a Yucca-size repository would have to be sited every decade. With need for such sites increasing, chances are slim that local objections will be respected. High-level waste could be coming to a repository near you.

The DOE's record to date of handling nuclear waste does not inspire confidence. Three of its six low-level waste sites have been closed due to unanticipated leaks of radioactive material. The Waste Isolation Pilot Plant (WIPP) that the DOE is attempting to establish in New Mexico for waste from the U.S. nuclear weapons program is also in trouble. Storage rooms that were expected to be dry contain brine and corrosive groundwater that could eat through steel waste drums and cause an explosion.

Besides safety problems for our generation, related to transporting waste to burial sites (trucks crash and boats sink from time to time), much concern relates to future generations, because the wastes remain toxic for so long. With a half-life of 24,400 years, plutonium is dangerous for a quarter of a million years (ten times the half-life). Imagine all the time between the ancient Egyptians and ourselves and multiply that by fifty.

Our knowledge of geology does not permit safe predictions for such periods of time. Ten thousand years ago there were active volcanos in what is now France; seven thousand years ago there was no English Channel; and five thousand years ago much of the North African desert was fertile. We are in no position to predict geological events 250,000 years in the future. A National Research Council (NRC) report in 1990 concludes that research during the 1970s and 1980s increased, rather than decreased, uncertainty.

The durability of geological formations is only part of the problem. Buried wastes can harm future generations who unknowingly dig them up. How can we warn people whose languages we cannot predict? Ulrike Fink doubts the possibility of passing on the required knowledge, since records of coal mine locations in Germany have been lost in just one hundred years. "[D]ata losses nowadays may take place even faster: Data files stored just twenty years ago on magnetic tape can't be read by modern computers anymore!"

The United States DOE is not even trying to protect *all* future people who could be harmed by radiation we are creating. Its goal is to protect people for ten thousand years, whereas our nuclear mastery of nature threatens people who will not exist for 250,000 years. Harming these people is obviously immoral. But the best isolation method so far developed, encasement in copper drums four inches thick, may last for only 100,000 years and is considered too expensive for general application.

This lamentable situation comes as no surprise to those responsible for promoting the growth of nuclear power in the United States. James B. Conant, former administrator of the project to build the first atomic weapons, warned of nuclear waste in 1951. A panel of the National Academy of Sciences (NAS) advised in 1960 that no new nuclear power plants be licensed until the waste issue is resolved. But respectable voices of caution were not heeded.

Problems of high-level nuclear waste disposal are detected more easily where countries allow some (although far from ideal) freedom of information and citizen participation. These include the United States, France, and Germany. Japan appears prepared to avoid such difficulties by exporting its high-level waste to China where, as we have seen, secrecy is the norm and citizen protest is illegal. Luke Onyekakeyah, from Nigeria, told the hearing that "an Italian company carried tons of waste and dumped it at a small port in southern Nigeria called Koko." He worries that after human slavery, political slavery, and economic slavery, countries of the Third World will be subject to nuclear waste slavery.

The logic is the same as that resulting in the U.S. focus on Yucca Mountain. The weakest financially and politically suffer for the improvidence of the wealthy and powerful. Thus, for example, the Taiwanese government is storing nuclear waste on Botel Tobago, officially called Orchid Island, home of the indigenous Yami people, who were neither informed nor consulted.

Similarly, in 1972 the Nixon administration suggested that areas where indigenous people live in the southwestern and upper midwestern United States be used in ways that would proliferate nuclear contamination. They were to be designated "National Sacrifice Areas," and would be effectively uninhabitable for at least 250,000 years. This would eliminate the land base, and so the very existence, of the Navajo, Northern Cheyenne, Lakota, Hopi, and most Pueblos. According to common definition, this is genocide.

It reminds me of a story about Mohandas Gandhi. When a reporter from the *London Times* asked him what he thought of Western civilization, Gandhi replied, "That would be a good idea."

The question remains whether there is sufficient justification for what our civilization is doing. We release radiation into the atmosphere through uranium mining; release radioactivity in various processes that prepare nuclear fuel; multiply the radiation millions of times in nuclear reactors; risk thousands upon thousands of deaths in Chernobyl-type accidents; and force the poorest and weakest people, and future generations, to bear disproportionate shares of exposure, and risk of exposure, to toxic radiation. The next chapter discusses the supposed benefits of nuclear power. Is it really cheap and plentiful?

Chapter 6

Nuclear Power
and Human Oppression

After hearing so many people testify at the World Uranium Hearing against the use of nuclear power, I was having difficulty understanding what its proponents could say in its favor. I have since found that industry advocates varied their arguments over the years. In the 1960s, nuclear power was supposed to be extremely inexpensive. When the price of nuclear power started to rise, it was seen in the 1970s as a way of getting along with less petroleum. When oil prices collapsed in the 1980s, nuclear energy was promoted as better than fossil fuels for addressing environmental problems, such as acid rain and global warming.

Industry advocates still claim that it is safe, clean, inexpensive, and plentiful. Some doubt has been cast in the preceding chapter on claims of safety and cleanliness. This chapter examines claims of economy and plenty before discussing safety issues not yet addressed. When this is done, the use of nuclear power to reduce global warming will seem like a bad joke.

Government Subsidies and Financial Failures

Initial impressions in the United States that nuclear power was, or could be, inexpensive were due largely to extraordinary government subsidies. In 1957 Congress passed the Price-Anderson Act that limited a utility's liability in case of an accident to $560 million. This was necessary for the initiation of civilian nuclear power because private insurance companies refused to provide full liability coverage for nuclear power plants as they do for other power plants (and for other industries). Even in the 1950s, as government planners waxed confident, insurance executives recognized they could be bankrupt by a Chernobyl-type disaster.

The insurance industry was correct. Many experts put the cost of the Chernobyl cleanup at $360 *billion*. In his testimony at the World Uranium Hearing, Vladimir Chernousenko, who headed the cleanup effort, maintained the cost to be more than $850 billion. The Price-Anderson Act subsidized the U.S. nuclear industry indirectly by shielding it from high liability claims.

More direct subsidies were given for uranium enrichment, foreign sales, and research and development of reactors and waste disposal methods. If all development costs were included in the price of electricity that nuclear utilities charge to customers, prices would rise by 50 percent. And this is not the half of it. Probably exceeding all these subsidies combined are tax breaks in the form of investment tax credits and accelerated depreciation of assets. Because nuclear power is so capital intensive, these tax breaks benefit nuclear power twice as much as competing fossil fuel plants.

Even with all these subsidies, in 1983 the cost of electricity coming from nuclear power plants scheduled to be completed in the mid-1980s was projected to be 65 percent higher than the cost of electricity from new coal-fired power plants. Using electricity from new nuclear plants for heating would be like using oil at $240 a barrel, which is more than ten times the actual price of oil.

In view of these costs, it is no surprise that nuclear power has been financially disastrous for many utility companies. The Washington Public Power Supply System (WPPSS—best pronounced "woops") used the resources of more than one hundred public utility companies in the Pacific Northwest. Between 1974 and 1981 cost estimates went from $4 billion to $24 billion, forcing cancellation of four of the five plants originally planned. By 1981 the system was already $8 billion in debt, leading to the largest default on bonds in U.S. history up to that time.

Since then, costs of construction have continued to rise, forcing cancellations of nuclear power plants elsewhere in the country. In fact, no plant ordered after 1974 has been completed and put into service in the United States. One hundred and eight such plants have been cancelled since the Three Mile Island problems in 1979. A spectacular case is the Shoreham nuclear plant in New York. By 1988, $5.3 billion had been spent on the plant, more than twenty times the original estimate. The state of New York finally bought the plant for just $1 to save the Long Island Lighting Company from bankruptcy. For reasons of safety, the state has agreed never to use the plant to generate nuclear power.

The United States is not the only country where nuclear power has proven economically hazardous. France has an ambitious nuclear power program known for efficiency in construction. On average, plants are built in France in only six years, compared to twelve years in the United States. Yet by 1991 the French state utility was $37 billion in the red—as much as Poland's entire foreign debt. France now exports 10 percent of the electricity it produces and justifies expansion of its capacity by projecting an export of 25 percent. But exports just add to the debt, because even the French cannot produce atomic energy as inexpensively as electricity is sold on the world market.

This is like the old joke about the merchant who increased business by selling items below his purchase price. Asked how he could make money this way, he replied, "By high volume."

Borrowing from Future Generations

As discouraging as they are for advocates of nuclear power, the costs discussed so far are gross underestimates. They do not include the cost of effective cleanup operations at uranium mines. The German government expects to spend $3.6 billion to clean up waste from just one Wismut mine in the former German Democratic Republic (East Germany). Others estimate that cleanup at three Wismut mines together could cost as much as $23 billion. Untold additional billions will be needed to reclaim all the other uranium mines in the world. None of this is figured into the already high price of electricity produced by nuclear power. Also not included is the cost in dollars, as well as in human terms, of an estimated forty-thousand radiation-induced deaths worldwide. These result from the mining needed each year to supply enough uranium to maintain nuclear power at its current level.

The burial of high-level wastes is another cost not yet adequately reflected in anyone's utility bill. Between 1983 and 1991 cost estimates for burying a ton of high-level waste rose 80 percent, with a single site holding 96,000 tons costing $36 billion. Because five times this amount of waste will come from reactors currently operating or being built, burial will eventually cost $160 or $170 billion, if no new plants are started. No permanent burial at this cost has begun because no technology has been accepted yet for such "inexpensive" disposal. Sweden's relatively safe approach, employing copper drums 4 inches thick, is more expensive.

One suggestion for handling high-level wastes is to reprocess them, as at Sellafield in England. High-level waste from reactors contains mostly uranium-238, but also plutonium and some remaining uranium-235. When isolated through reprocessing, the plutonium and uranium-235 can be used as fuel in reactors.

This sounds initially like virtuous recycling, like recycling aluminium cans. However, the cost of reprocessing is more than twice estimates for direct, permanent storage of wastes, and reprocessing increases the volume of wastes 160 times. This higher-volume, lower-level waste must then be dealt with.

Reprocessing also requires specialized facilities that do not exist near most nuclear power plants, where high-level waste is generated. Currently, only England and France have such facilities in large-scale, commercial operation. Plans for Japan's use of these facilities include shipping thirty tons of plutonium from England and France to Japan between 1992 and 2010, with about one ton in a single ship.

Anyone who has visited a beach on Florida's Atlantic coast, England's south coast, or any number of other places in the world, will notice tar deposited from oil tanker spills. After all recent experiences of jeopardy at sea, it is now proposed that large quantities of the most toxic substance in the world be put to sea. Besides ship wrecks, consider the possibilities of sabotage or simply accidents in loading or unloading. After a shipment of only one-quarter ton in 1984, the U.S. Defense Department, which provided escort for the ship, recommended no further sea shipments: "Even if the most careful precautions are observed," they concluded, "no one can guarantee the safety of the cargo from a security incident."

Besides this, plutonium isolated through reprocessing can be used to make nuclear weapons. Security considerations related to weapons are discussed later.

Low-level wastes and decommissioning power plants must be considered, too, when estimating the real cost of nuclear energy. Low-level waste includes already discussed tailings from mining and milling uranium. In addition, all parts of a power plant that have become irradiated in the process of nuclear power generation are low-level waste, even workers' clothing. The problem is that this material can spread its radiation contamination to whatever comes into contact with it. Because contact is necessary to move it about, clean it up, and bury it, most attempts to deal with low-level waste produce additional low-level waste.

Decommissioning is the process of shutting down a nuclear power plant, and making sure it is no longer hazardous, after it has generated electricity for its anticipated useful life, which is between thirty and forty years. For reasons just mentioned, dismantling the plant—carrying its radioactive materials to a suitable site for burial and burying it—multiplies considerably the amount of waste requiring burial. In forty years a typical reactor will produce 6,200 cubic meters of low-level waste, but dismantling is expected to produce an additional 15,480 cubic meters of such waste.

Technologies for dismantling are not fully developed yet, but estimates for decommissioning run as high as 50 percent of initial construction costs, which are themselves many times original projections. Again, these costs are hardly reflected in current utility rates.

A less expensive approach to decommissioning is entombment, basically covering the plant with concrete. But our best concrete has a useful life of five hundred years, and materials in decommissioned plants will be dangerous for one hundred thousand years. Entombment just shifts to future generations a lot of the costs associated with current power generation. If we feel too poor to clean up our own wastes, imagine how poor we would feel if we had to clean up the wastes of previous generations. Now just apply the golden rule.

The Scarcity of Uranium

Nuclear power is supposed to be plentiful as well as cheap, but supplies of rich uranium ore are limited. Poor quality ores require a lot of energy for uranium extraction, and much of this energy would have to come from fossil fuels that emit carbon dioxide and contribute to the greenhouse effect. Using uranium of poor quality for nuclear power could be worse for global warming than direct use of fossil fuels to produce electricity.

There are only six million tons of uranium ore rich enough to be used without worsening global warming. Even with currently uneconomical reprocessing, this would last only twenty-five to fifty years if all electricity was produced by nuclear power in plants like those used today. Current nuclear technology clearly is not a solution to the world's long-term energy needs.

The industry addresses this problem by advocating use of fast breeder reactors (FBRs), which could enable the same uranium resources to produce sixty times as much energy. Normal reactor fuel contains 97 percent uranium-238, which is not fissible, and only 3 percent fissible uranium-235. The bombardment of uranium-235 with neutrons to get them to break apart bombards the uranium-238 with neutrons as well. Some atoms of uranium-238 pick up an extra neutron and become plutonium-239, which, like uranium-

235, can be used as fuel in commercial reactors. In breeder reactors the conversion of uranium-238 to plutonium is facilitated by removing the moderator that slows down neutrons in conventional reactors. Fast neutrons leave the core, hit surrounding uranium-238, and produce plutonium.

This is inherently more dangerous than processes inside conventional reactors because more neutrons are flying around faster. This creates extra heat that requires a special coolant, usually liquid sodium, that may burn on contact with air and explode on contact with water. The world's first fast reactor suffered a core meltdown and almost exploded in 1955. The Fermi fast reactor near Detroit almost exploded in 1966. President Jimmy Carter, a nuclear engineer, recommended terminating the U.S. breeder reactor program, which Congress finally did in 1983. In 1991 Germany gave up development of its prototype breeder reactor. Due primarily to safety considerations, the French reactor Super-Phoenix has been mostly out of service since its completion in 1986.

Accidents at FBRs could have much more serious consequences than accidents at conventional reactors because FBRs contain much more plutonium.

Plutonium as a Military Threat

The worst aspect of any process that produces and/or isolates plutonium concerns security, as plutonium is the major ingredient in nuclear weapons. Plutonium is produced in current reactors and can be found in spent fuel, but this plutonium is not suitable for use in weapons without special reprocessing that isolates it. In principle, then, if reprocessing were controlled, nuclear power could be generated around the globe in conventional reactors without widespread availability of plutonium for nuclear weapons. It is the goal of the Non-Proliferation Treaty (NPT) of 1968 to realize this possibility.

Signatories must allow their nuclear facilities to be inspected to make sure that all nuclear materials are accounted for and weapons-grade materials are not being produced. In return, states lacking nuclear weapons receive help in the development of civilian nuclear power from those who do have such weapons.

This treaty does not guarantee nonproliferation. First, not every country interested in nuclear weapons initially signed the treaty. These include France and China, among states with nuclear weapons, and India, Israel, Argentina, Brazil, Pakistan, South Africa, and Spain, among nations who supposedly lacked such weapons when the treaty first came into effect. Second, a state that has signed the treaty and wishes to avoid inspection after receiving help in starting its (purportedly) civilian nuclear energy program, can simply withdraw from the treaty and refuse to be inspected. This was North Korea's threat in 1993 and 1994.

A third problem is that some of the fissible material coursing around the world (for peaceful use) can get lost. It was revealed in 1977, for example, that 200 tons of uranium disappeared at sea. It was probably diverted to Israel, a country now widely presumed to have nuclear weapons.

The fourth problem is even worse. The NPT permits countries to have the kinds of reprocessing plants that isolate plutonium. Because no accounting system is perfect, diversion of enough plutonium to make a few bombs each year could easily go undetected.

Biologist Raul Montenegro testified at the hearing that before construction was stopped, his native Argentina had completed 80 percent of a reprocessing plant designed to produce five tons of plutonium per year. The Argentinian government that started this project invaded the Falkland Islands and probably murdered many of its young dissidents. Its motives for building a reprocessing plant may have been military, since civilian, economic motives are ruled out for countries using little nuclear power. The same can be said of similar programs in Pakistan, South Africa, and North Korea, not to mention India, Israel, and Iraq, whose military goals seem clear.

Iraq's case is instructive. The NPT and the IAEA rules allowed Iraq to acquire the materials and reprocessing technology needed to produce and separate plutonium, as long as international inspection was allowed. But there are usually several months between inspections. Once all materials and facilities are in place, enough plutonium to make a bomb could be diverted and employed within a week.

Consider the implications of widespread use of FBRs designed to produce large quantities of plutonium in the process of producing electricity. If Iraq had been using an FBR, there would have been no way to prevent their acquisition of nuclear weapons, because relatively few, and quick, steps are needed to apply the FBR's material to military purposes.

The Global Warming Rationale

Consider now the proposal to use nuclear power to reduce global warming. Reducing global warming 20 to 30 percent in sixty years by generating electricity with nuclear energy instead of fossil fuels would require completing one nuclear power plant every one to three days during that period, and would cost $9,000 billion. Much of the money would have to come from the already debt-ridden Third World, because that is where nearly half of the plants would have to be located. This eighteenfold increase in nuclear capacity would produce corresponding increases in high-level waste for disposal and in power plants needing eventual decommissioning.

Just in case your disbelief counter is not yet ticking wildly, consider that most new power plants, whose construction would have to start immediately, would have to be FBRs; there is not otherwise enough uranium to provide such quantities of nuclear power. However, even industry advocates concede that FBRs will not be available for commercial use until the first or second decade of the next century, which is too late for this approach to global warming. Worse than this, it is not clear that such plants will ever be developed. As noted above, the United States and Germany are no longer even trying to develop such reactors.

Now turn off that noisy disbelief counter and imagine the situation if FBRs could be made to work, and there were about seven thousand of them in the world. At least three thousand would have to be in the Third World, where technological competence may be marginal and economic considerations may compromise safety. Reactors are still operating at Chernobyl, and others of the same design continue to be used elsewhere in the former Soviet Union (CIS), because these countries are too poor to forgo the immediate benefits of electricity generation. However, these countries are still better off financially and technologically than many in the Third World who would have to have inherently more dangerous FBRs if nuclear power is to reduce global warming.

Developing at least three thousand new plants in the Third World would require *thousands* of sea shipments of plutonium like those now underway between France and Japan. Enough said.

The worst problems concern the proliferation of nuclear weapons. Consider three things: 1) Nuclear weapons can be made relatively quickly and easily once the required plutonium is in hand; 2) when reloaded each year, a breeder reactor contains enough plutonium for hundreds of nuclear weapons; and 3) no fuel accounting system can preclude some undetected diversions. With three thousand breeder reactors in the Third World, any and every country that wanted them would have nuclear weapons.

Assuming that the Age of Aquarius does not miraculously appear, many Third World countries with large supplies of plutonium will be at odds with one another and with First World countries. Third World debt to the First World and to international banks will have increased due to borrowing needed to buy nuclear technology. Economic privation and a history of political instability are bound to produce some ruthless heads of state, like Pol Pot in Cambodia, Idi Amin in Uganda, and Saddam Hussein in Iraq. Then what will we do? It is obvious, because it is what we have already done. We will bomb them before they bomb us. Even using conventional weapons, however, bombing nuclear facilities may spread enough radiation to poison millions of people.

Another concern is that nongovernment groups will obtain nuclear weapons, with or without the help of governments. Imagine a world where criminals or terrorists could obtain nuclear materials by bribing a ship's captain or by winning the favor of sympathetic governments. Imagine if the group that bombed the federal building in Oklahoma City could credibly threaten to use a nuclear device next time. This is obviously unacceptable so we will do all we can to prevent it. Having created such danger, governments will curtail civil liberties, as President Clinton quickly suggested in the wake of the Oklahoma City bombing. People will not only accept increasing regimentation, but will applaud it and ask for more. They will see increasing searches and seizures as needed to prevent disaster, just as most of us (me included) applaud strict security measures at airports. Soon no search of a car, a home, or a person will be considered unreasonable.

Restrictions on democratic processes will be accepted as necessary to prevent the disaster that could result from an ill-informed public electing officials who are "soft on terrorism," or who, like Saddam Hussein, may threaten neighbors with nuclear annihilation. In 1993 and 1994 democratic processes were on hold in Algeria because dangerous people, Islamic fundamentalists, won the last election. This denial of power would seem all the more reasonable if power included access to nuclear weapons.

Having created a Hobbesian state of nature, we will have justified a Hobbesian Leviathan to whom all power is surrendered. This could be the United Nation's Security Council, led by the United States and a few partners.

This is the classic pattern:

1. People consider themselves separate from nature.

2. They believe that nature does not provide enough to eliminate scarcity, so . . .

3. They develop fossil fuel energy to manipulate nature more effectively (to reduce scarcity), but fossil fuels risk global warming. Instead of questioning the initial project of controlling nature so completely, the problem is addressed with another technology, nuclear power, that promises even more power over nature. This technology is inherently toxic.

4. Centralized government bureaucracies rule from the beginning, both to protect many people from the worst consequences of atomic power and to protect decisionmakers from adverse public reactions if real costs are exposed. Many decisionmakers have a personal and professional stake in nuclear power, and propose massive construction of FBRs to control global warming.

5. From the beginning, governments put the costs of nuclear power largely on indigenous people, poor people, those living near nuclear power plants and waste facilities, and future generations—all relatively unable to protect themselves. Using FBRs makes danger so complete that it justifies a global police state. Bureaucrats of a world government decide whose home is searched and whose neighborhood is invaded with nuclear waste.

This is not a prediction. While I am no great optimist about human prospects in general, even I expect us to be wise enough to avoid worldwide totalitarianism. I expect us to foresee the danger and reject the nuclear path. My point is that these results do follow from the logic of addressing global warming with fission technology, that is, FBRs. Remember the Chinese proverb: If we do not change direction we are likely to end up where we are headed.

The Gulf War

In case it does not seem that this is where we are headed, consider what has occurred already in response to dangers associated with commercialism and nuclear proliferation. Consider the Gulf War in historical perspective.

As we have seen, commercialism and industrialism require secure raw materials, markets, and travel routes, prompting conquests around the world by major industrial powers. Britain, France, Japan, and the United States illustrate the point. Britain and France assumed preponderant power in the Middle East following World War I, as Mideast oil, travel routes, and strategic position (relative to the new Soviet Union) became increasingly important.

Imperial powers have known at least since Roman times that local administration facilitates management of territory under imperial control. Western powers carved up the Middle East into nation-states and installed friendly leaders with enough power to maintain order and facilitate Western access to oil, markets, and trading routes. Iraq is one of these states. The fact that there is an Iraq to fear and fight is an artifact of Western industrial-commercial interests.

The central government in Iraq needed to be strong to accomplish all that Western powers desired of it. Iraq includes in the north Kurds with their own aspirations of self-rule. Western interests oppose these aspirations because their realization would upset Turkey, a Western ally. Eastern Turkey has many Kurds who would like to join Kurds in Iraq to form a Kurdish state. Turkey opposes the loss of territory that would result from the creation of such a state, so part of Iraq's job was to repress its Kurds to forestall demands for secession by Turkish Kurds. (Because the Gulf War weakened Iraq, Turkey has had to act for itself in this matter, sending 35,000 troops into northern Iraq in March 1995.)

A similar task concerns Shiites in southern Iraq. They share religious doctrines with the majority of people in Iran, which is already a populous and powerful country. Iranian assimilation of Iraqi Shiites would upset the power balance in the Mideast. Iran could become powerful enough to defy Western interests and threaten Israel even more than at present, so repression of Shiites in the south is also part of Iraq's job.

Repression requires power, which helps to explain why Western interests supported increasing power concentrations in the ruling party and in its chief, Saddam Hussein. Hussein was supported by the United States, not despite his violations of human rights in the north and south, but precisely because of those violations, which supported U.S. interests.

Western support for the Iraqi regime included provision of nuclear technology. This supported Western interests in two ways. By giving Iraq what it wanted, Western countries perpetuated and strengthened a beneficial relationship. Second, the sale of nuclear technology anywhere aids nuclear industries in Western countries by developing markets for their products, thus providing jobs in Western countries and justifying capital investments.

Western countries created a dangerous situation in Iraq. They installed a repressive government and by placing nuclear technologies in its hands, put nuclear weapons within its reach. (Fortunately, the nuclear technologies did not include FBRs, which would have made nuclear weapons available almost immediately). Having installed a vicious guard dog to terrorize everyone within a fenced-in area, having provided the means for it to jump the fence, and having approved initial fence-jumping (the attack on Iran), we finally decided that the attack on Kuwait was going too far.

What was the Gulf War all about? Restoring a certain family to power in the tiny country of Kuwait was not particularly important in the global scale of things. Vindicating the principle of national sovereignty against foreign invasion could not be very important to the United States. We approved Iraq's invasion of Iran and recently invaded Panama and Grenada ourselves. Protecting Mideast oil supplies was not a sufficient motive for attacking Iraq, because oil supplies could have been guaranteed by sending troops to Saudi Arabia to guard against an Iraqi invasion.

The only strong reason for actually attacking Iraq was to prevent them from developing nuclear weapons, which, given their access to technologies used in the production of nuclear power, would be only a matter of time. The right to inspect Iraqi installations to assure the absence of nuclear weapons production was a major element in the agreement formally bringing hostilities to an end. The threat of nuclear proliferation has been the focus of the United Nations' attention since then.

(The same security concern dominated U.S. thinking about North Korea in 1993 and 1994. In November 1993 a nationwide poll revealed that 59 percent of Americans favored military intervention against North Korea to deprive them of nuclear weapons, but only 31 percent would intervene if the North merely invaded the South.)

Thus, the Gulf War follows a familiar pattern. People create jeopardy in the course of mastering nature in the human interest. A central power is needed to suppress people to reduce the danger so created. As nuclear power is the greatest so far developed, and creates the greatest danger, the greatest repression from the most centralized power source is considered justified. The United Nations' Security Council, the most centralized source of "legitimate" power, authorized a war in which about 250,000 Iraqis, mostly civilians, were killed. Two hundred fifty thousand people paid with their lives for the insecurities created by fission technologies.

Rejecting Responsibility

The most shocking thing is that, outside the Arab world, most people considered the war perfectly justified. The part played by Western powers in the creation of a dangerous situation is almost ignored. The war is blamed on Saddam Hussein without considering why such a person was in power.

This is not an isolated case. As we have seen, the ideology of acquiring power over nature in the human interest also has been used to rob native peoples of their land and way of life. This continues today in Canada, Siberia, Brazil, and Tibet, to name but a few. The ideology is used to uproot peasants from fertile areas previously used for subsistence agriculture. This is a major reason why 12 to 15 million children die each year due to illnesses stemming from malnutrition. A common reaction in the United States to the plight of destitute natives and starving peasants is to blame them for having too many children. We conveniently ignore how commercial-industrial intrusions (to benefit humanity) have disrupted previously adaptive traditions.

We protect our moral sensibilities also by remaining conveniently ignorant. How many Americans realize that children are starving in fertile areas of Central America because the land cultivated by their ancestors is now used to raise beef for hamburgers sold in the United States?

Finally, we implicitly deny the humanity of those whose suffering cannot be ignored entirely. Discussions of casualties in the Gulf War usually focus on deaths among allied troops, as if Iraqis are not people. "How many people died in the war?" I heard it asked. "One hundred eighty-six or so, but some of that was by 'friendly fire.'" That such a response was given and accepted widely in the United States suggests how far in the age of nuclear jeopardy we fall short of our own ideal of respect for human beings. Another indication of the lack of respect for human life is popular acceptance of starvation in "overpopulated" countries while there is plenty of food in the world to feed everyone.

In sum, we morally neutralize the status quo by passively accepting lame excuses, such as blaming the victims; by remaining conveniently ignorant of other people's misery; and by implicitly denying the humanity of those socially, culturally, and/or physically distant.

European Jews were blamed similarly for their predicament. They were, for example, criticized by Nazis for their failure to put down roots and work the land. Nazis conveniently forgot that Jews were disallowed land ownership in most parts of Central Europe until 1848. Average German citizens, even those living just outside the walls of Dachau, claimed ignorance of what was happening to Jews inside, and the full humanity of Jews was put in question.

The concept of humanity as separate from, and superior to, nature, and the related project of subduing nature in the human interest, have brought us to this. We are so distant from one another and from our moral ideals that fundamental change is required. It is time to think the unthinkable. We need to consider how life could be if humans acted as plain members and citizens of the ecosystems they inhabit rather than as lords and masters. This is the larger context for rejecting nuclear power.

Chapter 7

Indigenous Peace and Prosperity

Why Discuss Indigenous Cultures?

The World Uranium Hearing was designed largely to publicize problems that many indigenous people face as a result of our culture's use of uranium technologies. But the testimony of these people, and of scientific experts, displayed a larger pattern. Our culture tends to oppress people in the process of subduing nature in the supposed human interest.

This implies that much of what I was taught to view positively as progress must be reevaluated and reclassified in light of our culture's commitment to human rights. Few commercial, scientific, or industrial developments can be considered progress any longer because they increase human oppression.

I cannot leave matters here, however, because I cannot give up on promoting progress, *real* progress. The hearing testimony by native people about their cultures suggested that there are cultures where human beings are less oppressed than they are by, and within, our culture. The present chapter, concentrating primarily on foraging (hunter-gatherer) societies, describes viable ways of life that involve much less human oppression than our own. Real progress may come through learning from these people.

The next chapter concerns a social conception of the nonhuman environment and a reverential attitude toward nature that are common in world views of indigenous people, especially foragers. Foragers do not consider humanity at odds with nature and so do not try to manipulate nature maximally in the human interest. Their perspective resembles Aldo Leopold's injunction in "The Land Ethic" to be plain members and citizens of our biotic communities.

These two chapters do not maintain that societies in which there is less human oppression and nature is valued for itself are typical of indigenous societies, much less that all indigenous societies are like this. The claim is merely that there are *some* societies that fit this description, and they all happen to be indigenous. Their existence supports the thesis that *in societies where there is much less human oppression than in ours, nature is typically respected as valuable in and for itself, and people are not trying to overpower nature for human benefit.*

This is not to say that people are treated better wherever nature is valued for itself. There are murderous people who value nature for itself. But where there is relatively little human oppression, people tend to revere or respect nature, and do not try to dominate it.

Even if there are such indigenous societies, how does that affect us? We certainly cannot become indigenous ourselves. Whether studying these cultures can prompt positive changes in our culture is an open question.

Stateless, Egalitarian Indigenous People

The native people that give the greatest reason for hope are traditionally stateless. They do not live in societies whose traditional culture is shaped by the presence of a state. According to sociologist Max Weber, the state is that organization in society claiming to determine how physical force is used. This does not mean that the state forbids all others the use of physical force; parents may spank children and linebackers may tackle running backs. But the state sets limits. Hitting a child with a bat in the head would be child abuse, and tackling the running back in the parking lot is assault.

The state can control the use of physical force in society only by having superior force at its disposal, enabling it to force compliance and/or punish deviance. The state can punish law breakers because it is more powerful than they are. When this condition is not met, there is often widespread lawlessness. Bands of armed militia may rob innocent people with impunity because there is no state strong enough to subdue them, as was reported in many parts of Somalia in the early 1990s.

Brutal lawlessness in the absence of a state is just what Hobbes would expect. He assumed that people are naturally so individually selfish that they would not cooperate with other people, or refrain from killing others, except out of fear for their own safety. Violence is curtailed only when there is a central organization dedicated to maintaining peace that wields more force than any other individuals or groups. Hobbes calls this the Leviathan, Weber calls it the state.

Much indigenous life belies Hobbesian expectations. Cooperation and sharing are common in many stateless tribal societies, where resources needed in production are treated as a commons. Traditional foragers, for example, share a geographic area where food is gathered, water is found, and animals are hunted. Most equipment for these activities is available to everyone because most people know how to make equipment for themselves from resources, such as egg shells, sticks, and animals, found in the common geographic area.

Slash-and-burn horticulture is another production method used by indigenous people. Within a relatively large area, patches of forest are cut down and burned to clear small areas where food is grown for one or more seasons. When yields decline, the group repeats the process elsewhere in the forest, which is held as a commons.

Herders (pastoralists) often exist within states, but their cultures are similar in important respects to those of many stateless indigenous people. Whereas animals are usually owned by individual families, there are typically restrictions on their sale and transfer. Owners do not have all the freedom to dispose of what they own as in our society, so ownership does not carry as much power or importance among them as among us. Also, the land area used to feed the herds is not owned individually at all. It is held by the group as a commons, like the common geographic area used by foragers. Their use of a commons area and their restrictions on commercial exchanges make herding cultures similar to those of stateless indigenous people.

These are only three types of commons-oriented economies. They share several features. First, there are rules about who can use what part of the commons when, but these rules are backed by informal methods of social control, such as social disapproval and personal shame, rather than by police. People in our society who think that social disapproval, personal shame, and socially induced tastes cannot effectively control behavior should consider what keeps most men in the United States from wearing a dress.

Rules of the commons may dictate that no family in a slash-and-burn economy plant more than a certain amount of land. In a foraging society they may dictate that meat from killed animals be shared widely, and that scarce water resources be shared for the asking. But they may also dictate that guests contribute food after only a few day's stay and honor hints that their stay has become burdensome.

Second, no one in traditional, commons-oriented societies is driven to absolute destitution while others have enough. Absolutely destitute people may become desperate, ignoring social rules of the commons. Because traditional indigenous people are stateless, they lack formal, authorized, physically coercive methods of controlling social deviance. So they forestall deviance by insuring that every nondeviant can live a decent life.

Corresponding to the provision of material essentials for everyone is the absence of great accumulation by relatively few people. Some people have more than others, certainly, and this is often socially important to everyone. But indefinitely large accumulations do not exist. There are often social rules to insure redistribution to others, especially those most in need, of any great excess. This is the flip side of insuring a materially tolerable life for all.

An example is provided by the Ju/'hoansi (pronounced "Jut-wasi," where "j" is pronounced as in the French words "je" and "ja-mais"). These traditional foraging people of southwest Africa, whom anthropologists formerly referred to as !Kung, value meat highly. Nevertheless, the hunter does not own the meat taken from the animal he killed. Rather, it is owned by whoever owns the arrow that was used. Because arrows are traded, the "owner" may not be a hunter at all, but a woman (whose gendered roles do not include hunting). Ownership, however, does not permit exclusive possession or use. The owner has the duty of distributing the meat throughout the group, which may include between ten and one hundred people. Even if meat could be stored for months, it would be considered uncivil for someone to keep a large supply while others had none.

The Gabra are a nomadic herding people who travel with camels in search of food in the forbidding Chalbi Desert in East Africa. Like the Ju/'hoansi, the Gabra expect generosity, maintaining that "a poor man shames us all."

The same view was expressed at the World Uranium Hearing by Galsan Tchinag, a witness from Mongolia belonging to the indigenous people who call themselves Tuwina. He told us, "The bounty of the hunt is shared, not just among the hunters who took part in the hunt, but among everybody who appears and asks. And so it is with riches and abilities."

Matters are different where there is a state. The state is a force strong enough to oppose those who would steal from more affluent members of society. Thieves are deterred by police or isolated from society in prisons when they are caught so that inequalities can be maintained despite poor people's needs or jealousies.

The police may be compared to dams that enable still water to exist in close proximity at different elevations. This is done by exerting constant pressure on the water. Take the pressure away and water seeks its own level; the differences in elevation are eliminated. If the differences in elevation are great and a lot of water is backed up behind the dam, removing the dam can create a catastrophic flood. Similarly, take away police in our society and people will tend to even out the wealth by stealing from those more affluent than themselves. Property crimes are known to increase when police are on strike or otherwise incapacitated. Property crimes increase also (other things being equal) when differences in wealth increase. The greater the gap between the rich and poor in society, the more crime there is in the absence of police protection, and the more police and other manifestations of state power are needed to maintain relative tranquility. Probably recognizing this, the Reagan administration conjoined cutbacks in services to the poor in the early 1980s with increased expenditures on law enforcement and incarceration facilities. Reverting to the dam analogy, the greater the elevation difference between proximate waters, the stronger must be the dam that separates and maintains the difference between them.

Followers of Hobbes point to lawlessness and violence in the absence of police protection as evidence that states are absolutely essential for human survival and well-being. They assume that stateless indigenous people have miserable lives, but this assumption is not justified. The sudden removal of a dam can be catastrophic when the dam has been used to maintain unequal water elevations. The more unequal the water elevations, the worse the flood. But no similar catastrophe need attend failure to construct a dam in the first place. Indigenous social environments are like water where no dam has been constructed. There is relative equality and the flow of material and social goods is relatively smooth and gentle.

Statelessness and Violence

Some people claim that states serve not just to permit material and social inequality, but also to control violence that is endemic to the human condition. Continuing the comparison with dams, a stateless society is compared to a river that floods its banks regularly and uncontrollably. In this situation, dams and dikes are needed because problems exist when they are absent, not just when they are suddenly removed. Similarly, people are naturally violent and socially disruptive. Just as the river needs dams and dikes, people need a state.

There is no doubt that people are sometimes violent toward one another and that state intervention can often reduce *locally* the incidence of such violence. Among the Ju/'hoansi, sexual jealousy, infidelity, and disputes about meat distribution are major underlying causes of serious violence. There were twenty-two killings between 1920 and 1955, which approaches the per capita homicide rate of dangerous American cities in the early 1970s. Because the Ju/'hoansi seemed dangerous to their neighbors, in 1955 the Botswana government appointed a headman who administered customary law for the next twenty-five years. There were only two homicides in the area of his administration. All of these statistics suggest that people are inherently violent and that states are needed to provide overriding force to reduce violence; however, these appearances are deceptive.

First, the Ju/'hoansi lacked Western medical techniques, such as surgery under sterile conditions, that can reduce an otherwise lethal injury to a temporary inconvenience. Their homicides include deaths from injuries that would be inconsequential in Detroit.

More important, the suppression of violence at one level of social organization probably leads periodically to outbreaks of considerably more violence at a more inclusive level of social integration. This extra violence is sometimes classified as war, so the killings are not called homicides, but they are deliberate killings of some people by others all the same. Anthropologist Richard Lee gives the following example:

> In the nineteenth century the Botswana chiefdom imposed its order on the band-level San hunters in Eastern Botswana, only to wage intertribal warfare on a much larger scale against neighboring chiefdoms such as the Matebele and the Kalanga-Shona. Then at the end of the nineteenth century, the British industrial state brought the Pax Britannica to the warring chiefdoms of Southern Africa. But a gener-

ation later, the British mobilized thousands of Tswana warriors' sons to fight in the Mediterranean theater against the German and Italian states.

In general, violence suppressed at one level of social organization probably does not disappear. It is simply externalized from that level to a more inclusive level where it costs more lives, so states may promote more interpersonal violence than they prevent. (Again, an analogy can be made to hydrology. Building dams and dikes to control periodic floods can result in worse floods in the long run. Dikes do not allow use of floodplains, so rivers swell during heavy rains much more than they otherwise would. When rains are extraordinary rivers spill over or destroy dikes, and the flood is much worse than it would have been with no flood control in the first place. This is what happened in the Upper Mississippi River system in 1993.)

None of this should surprise readers of the preceding chapters. Controlling people in greater numbers over larger territories is required by commercialism and industrialism. Nation-states must control hostilities among diverse groups of people into whose hands technology places ever more lethal weapons. States use powerful technologies of destruction to fulfill these functions and are organized bureaucratically to foster efficiency and the execution of duty without personal feelings. Technology, social hierarchies, and bureaucracy combine to distance state officials from people, domestic and foreign, affected by their actions. This tends to mute natural empathy with such people, reducing inhibitions to violence.

The Semai, indigenous horticultural people of Malaya known for nonviolence, illustrate how modern conditions can foster violence. To fight communist insurgency in the 1950s, British colonial rulers of Malaya recruited Semai to join the army. Not imagining that they would be called upon to kill, the Semai thought their job was just to tend weeds and cut grass. But, removed from their nonviolent social context, they became taken up by what they called "blood drunkenness" and were among the most violent soldiers. Upon returning to their own society they were as gentle as ever and could not account for their violent behavior.

Because natural, empathetic restraints on human aggression are weakened at the nation-state level, it is no surprise that interpersonal violence thrives there, and that peaceful people can be violent in that context and that context alone.

Of course, not all indigenous people are as peaceful as the Ju/'hoansi and Semai. The Yanomamö, for example, are Amazonian horticulturalists known for being warlike. My point is not that all stateless, indigenous people are less violent than us, but that statelessness does not necessarily produce the increased violence predicted by Hobbes. Further, where traditionally egalitarian, nonviolent people are concerned, state authority seems to promote an increase in overall interpersonal violence.

Food Abundance and Population Control

We are taught that government is necessary to uphold the economic system of private property, which is required for people to meet the material needs of life. Hobbes said that life would be poor without the state, and Adam Smith said that scarcity is endemic to the human condition. Without a state to protect private property, it would seem that people would be relatively equal, as noted earlier, but pitifully poor, perhaps even starving. However, the Ju/'hoansi have a varied diet containing sufficient calories and nutrients. Although many foods are available only seasonally, there is no season when people tend to lose weight for lack of food. The same has been shown of other foraging people, and anthropologists now accept the generalization that the foraging way of life meets people's biological needs.

Perhaps more surprising, the Ju/'hoansi and other foraging people meet their needs with relatively little work effort. When all forms of work are considered, including tool construction, maintenance, and house work, men work about 44.5 hours per week and women 40.1. In our society, people typically work 40 or more hours per week for money, and then work an equivalent amount doing laundry, preparing meals, shopping for food, caring for children, and so forth. At least some foraging people meet all their needs with much less work than we typically do.

The benefits of foraging require that human population concentrations be low enough for the common land area, where people hunt and gather, to supply enough for everyone. Thomas Malthus would not have expected this situation to persist. Writing at the dawn of the Industrial Revolution, he maintained that population tends to increase faster than food resources, resulting in periodic overpopulation, food shortages, and starvation. This view supports the commercial and industrial belief that scarcity is endemic to the human condition.

Foraging people do not display the Malthusian tendency to overpopulation and periodic starvation. We have seen already that their nutrition is more than adequate. In addition, they seem to have stable populations over long periods of time. This may be due in part to early deaths resulting from lack of sophisticated medical care and some practice of infanticide. Probably most important is intelligent control of the birthrate.

Foragers lead a nomadic life, shifting camp sites several times during the year to obtain food in season. Lacking means of transportation other than walking, small children must be carried during seasonal treks. This gives adults, especially women, who do most of the child-carrying, incentive to space births.

Birth-spacing may be accomplished in part by vigorous, long-term nursing of children. Although the reader is not encouraged to try this at home, nursing may inhibit the conception of an additional child. The widespread use in our culture of pacifiers and bottle feeding may increase overall birth rates by reducing the ovulation suppression effects of nursing.

Foragers are not the only indigenous people who control birth rates. Some other indigenous societies reduce birth rates through marital arrangements. In the Ladakh region of the Himalayas several brothers marry the same woman, while other women become nuns and bear no children. Elsewhere, cultural norms specify late initial marriages, and include restrictions on widows remarrying.

The fact that indigenous people can control their populations is implicit in the long-term existence of indigenous, nomadic herders. Because their stock of capital goods, the herd, is also food, failure to control population size would result in eating the herd to avoid immediate starvation, which would end the pastoral way of life. Because this does not happen, effective population control exists among herders.

The ability of indigenous people to control population size is underscored by the quick erosion of that ability when indigenous life is disturbed by commercial, and especially by industrial, cultures. The Ju/'hoansi who abandoned nomadic foraging for sedentary lives quickly doubled their fertility rates. As Britain became increasingly a commercial, and then an industrial, society, its population increased, going from around 6 million in 1700, and 6.5 million in 1750, to 9 million by 1800, and 17 million by 1850. A similar population explosion accompanied British rule in India. India's population was stable for two centuries at 100 million people, then rose to 130 million in 1845, 175 million in 1855, 194 million in 1867, and 255 million in 1871.

Because Thomas Malthus was writing in Britain at a time of significant population increase, it is no surprise that he thought such increase natural for human beings. Thinkers often incorrectly assume that their own, local circumstances are keys to reality in general. As noted in Chapter Three, Darwin's explanation of evolutionary change must be taken with a grain of salt for this reason, as it reflects the industrial economy in which he lived.

Additional doubt attends Darwin's theory because it depends on Malthus's views. Darwin supposed that individuals must compete for vital resources because the Malthusian tendency to overpopulate makes resources inadequate to meet everyone's needs. If humans do not tend to overpopulate, Darwinian competition and natural selection do not apply to human beings.

Malthus's theory of natural human overpopulation crumbles in light of current evidence that, ironically, the scarcity commercialism and industrialism are supposed to combat is largely their product. Populations tend to increase with commercial and industrial advance and this makes food scarce. There are more people starving today than ever before.

Poverty and Exchange

While many indigenous people have relative equality, sufficiency, and peace, their lives are materially simple compared to ours. Their self-sufficiency and leisure rest on the simplicity of their material wants and needs. Many people might equate such simplicity with poverty.

But material possessions are created and have great cultural importance as items of exchange in many indigenous societies. The point of most exchange is not the acquisition of needed items, however, since people already have these. The point is creation and maintenance of social bonds through obligation.

Where there is no state it is harder for people to settle disputes through appeal to authority. Because strong interpersonal animosity may precipitate violence that is hard to contain, people typically spend considerable time cultivating personal relationships to forestall dangerous animosities. Gift exchange is part of this process. When a gift is given, the recipient's obligation to reciprocate creates a social bond.

Because creating and maintaining social bonds is the goal, immediate return of an equivalent gift would defeat the purpose of exchange, as it would terminate obligation between the parties. So people usually delay reciprocity in order to alternate between being owed and being under an obligation. Among the Ju/'hoansi, therefore, when two people exchange gifts on a given occasion, the exchange involves two half exchanges. One person is giving a gift in return for what she received on a previous occasion, while the other is giving a gift that will be reciprocated in the future as part of the next exchange.

Another effect of concentration on maintaining social bonds is that Ju/'hoansi do not keep forever what they have been given. Instead, valued items are expected to make their way around to many people over time. There is no place for hoarding of any kind.

This is just the opposite of what people in our society seek in commercial exchanges, where people seek products, not friends. We aspire to personal, long-term accumulation of valued items. We (usually) seek simultaneous giving (where possible), and try to receive at least as much as we give. We counsel against indebtedness ("Neither a borrower nor a lender be"), although we do use credit cards when we "need" products right away.

However, noncommercial exchanges, designed to create and maintain relationships rather than acquire products, persist among us as among tribal people. We exchange birthday and Christmas presents, for example, even though people could more easily and reliably get the items they really want if everyone bought for themselves. Having people over to dinner is another example. It often initiates or strengthens social ties by creating an obligation to reciprocate. Clothing is sometimes used as well. During their teen and college years, two of my three daughters exchanged clothing extensively with friends. All had sufficient attire (to say the least!), so the exchanges were probably more to cement relationships than to improve physical appearance.

Industrial Poverty

This kind of noncommercial exchange is not the predominant one in our society. We emphasize commercial exchanges that make our lives, paradoxically, both materially richer and more impoverished.

In order for people to engage willingly in commercial exchange, each has to have something that the other wants. A commercial economy grows only when wants increase, keeping demand high, and production increases, keeping supply abreast of demand. Industrial manufacture boosts supplies, while advertising, social prestige, and several ways of creating needs increase demand.

This system spurs hard work, technological innovation, and increasing appropriation of the earth's resources for human ends. The result is material riches that indigenous societies cannot attain—televisions, radios, printing presses, cars, planes, computers, and running hot water.

But the result is also poverty. We saw earlier how commerce and industry create scarcity, and even starvation, for have-nots, especially in the Third World. The point here is that even the haves experience poverty among their riches.

It is no accident that people in the United States often feel poor, even those whose income is in the top 20 percent. The system would not work unless people were dissatisfied. They must have unmet needs to motivate them to work hard for money, and to exchange that money for goods and services supplied by others.

Pervasive inducements to consumption in our society include physical and social structures that result from previous consumption. Because cars are our primary mode of transportation in the United States, the physical structure of most communities reflects use of the car. When residential, work, and shopping areas are far apart, and there is little public transportation, cars are a requirement, not an option.

Genuine needs are created also by evolving social expectations. At one time people could apply for jobs with typed resumes. Now, any resume that does not reflect use of word processing (different fonts for variety in size, shape, and boldness of letters) may not be taken seriously.

The complexity of goods and services has motivated governments to require increased training and credentials before people can engage in certain professions. People need money for training to become doctors and lawyers.

Advertising creates new "needs" (in a society that does not distinguish wants from needs). Advertisements introduce people to desirable products that provide enjoyment, save time, foster success, avert danger, et cetera. The more subtle advertising message is that people lacking the product should feel deprived. Others will notice their deprivation and think less of them. More important, they will, and should, think less of themselves.

The items required to avoid such distress vary with time and with social and economic class. Levels of luxury and "performance" generally required in cars (to avoid feelings of shame and deprivation) increase yearly, but at the moment (I think) only relatively wealthy people feel the genuine need for a Mercedes. Times change, however. At one time only wealthy people felt they needed a color television instead of a black and white. Now, almost everyone has one. Soon, rich people may require high definition television. When the rest of us join them, they will move on to feel deprived without personal helicopters, twenty-feet-long bathtubs, or something else.

Our economic system thus frustrates genuine self-fulfillment, even though such fulfillment remains one of our ideals. Psychologist Abraham Maslow maintained that people have a hierarchy of needs, of which *self-actualization* is the fifth, and highest. It is the desire to realize all of one's capabilities, and involves the most profound self-fulfillment. Before that need can be met, however, people must meet needs for 1) *physiological* requirements (food, shelter, et cetera), 2) *safety*, 3) *love* (affection, acceptance, and companionship), and 4) *self-esteem* (which comes from earning the praise of others for independent action).

Self-actualized people are beyond the (love or self-esteem) need of social approval. They seek to develop their own capacities, rather than purchase mass-produced, prepackaged versions of fulfillment. For the most part, they are fulfilled by what they *do,* not by what they *own.*

Self-actualized people are likely to be poor customers. Some may spend a great deal because they are self-actualized by sailing large yachts or driving fast cars, but most, for reasons of taste or finances, will do things like read books, write poetry, paint, go bowling, or play the piano. Because they seek little praise from others, they are not easily influenced by advertising or peer pressure to follow fashion or change pursuits (which requires new purchases). In general, insecure individuals who seek approval from others, and change pursuits often, are better customers.

In a commercial culture, then, Maslow's hierarchy of needs is overshadowed by Wenz's "lowerarchy of worry." Once people fulfill a need and can stop worrying about it, a less inherently pressing need takes its place as the focus of worry. Once people have healthy teeth, they can worry about their teeth being perfectly straight or pearly white. Once people have televisions, they can worry about screen size or high definition, feeling genuine deprivation and low self-esteem if they cannot "move up."

Because there is no end to this, there is no lowest member in the lowerarchy of worry. If commercial society is supported by the lowerarchy of worry, it is worry all the way down.

The U.S. budget crisis reflects this. It is brought on largely by people's unwillingness to pay more taxes. They are unwilling because they feel poor already. They have many unmet wants (which cannot be distinguished from needs) whose fulfillment is cause for concern. This kind of concern must be common even among rich people, as most would resist an increase of their taxes to 60 percent of earned income. Thus, U.S. budget problems are just a reflection of the sense of dissatisfaction, worry, and poverty needed for the economy to grow.

The conventional assumption is that growing economies reduce poverty, but the experience of U.S. charitable organizations during the 1994 Christmas season belies this assumption. Food donations for poor people were smaller, and the need for them greater, than during the 1982–83 recession. But in 1994 the economy grew fast, unemployment fell to 5.6 percent, and inflation was low. These are ideal conditions according to conventional economics. The aggravation of poverty and degradation of means to address it illustrate that a growing economy can frustrate more than serve genuine human needs. Because the economy grows by convincing people that they need more than they have already, a growing economy spells growing selfishness among society's haves and a decreasing willingness to help have-nots.

There is profound truth in Will Rogers's quip during the Depression that Americans are the first people to go to the poor house in an automobile. It points to the fact that poverty has little to do with the number, variety, and power of the material possessions at one's disposal. Much more important is the relationship between what one has and what is socially expected. People in *Little House on the Prairie* and other accounts of nineteenth-century rural life lacked television sets and video cassette recorders, but they were not poor on that account. Yet many people today with television sets and video cassette recorders are in poverty because society is arranged for people to have more. Being relatively poor in a society that values material wealth, these people are genuinely unhappy.

This unhappiness is not some unexpected by-product of our economy. It is endemic, because it is the factor counted on to motivate people to compete, achieve, produce, and further material progress. The ironic result is that our way of life, taken as a whole, fails on the very measure that we take to be most important, that is, individual preference satisfaction, when compared with indigenous ways of life. Because the latter do not rely on dissatisfaction, more people are allowed to be satisfied. Satisfied people can afford to be generous to those in need, so there is generally more sharing and less poverty.

In sum, there are societies where people live reasonably satisfying lives and human oppression is less than among us. I turn now to the cultural representation of nature in indigenous societies to show that they typically value nature for itself. This suggests the possibility that if we learn to value nature for itself, as the land ethic suggests, we may be able to reduce human oppression among us.

Chapter 8

Indigenous World Views

Expressions of indigenous cultures were frequent at the World Uranium Hearing. Besides dress and language, indigenous people used song and prayer to convey their cultures' riches. The Monday afternoon session began, for example, with a prayer sung in the Hopi language. The present chapter discusses world views typical of tribal people and exposes some fundamental differences between their cultures and ours. It shows that indigenous societies where people are less oppressed than in ours have cultures in which nature is valued for itself.

Natural Sufficiency and Cyclical Time

A world view is a general account of what exists in the world, how the world's constituents interact, and what is valuable (worth preserving or attaining). Indigenous world views typically start with an assumption of plenty: When traditional techniques are used, nature supplies all that people need to lead decent lives. Some foraging people cannot believe that anyone ever dies of hunger.

This assumption of plenty coheres with statelessness. Without a state to keep people in line by force, indigenous people must get along with one another. No one can be allowed to fall into complete destitution and hopelessness, as this may provoke disruptive behavior that society could not handle. Successful population control enables traditional tribal people to continue meeting everyone's needs.

For the same reason there must be relative equality, and personal success must be common. Failure brings social disaffection leading to unmanageable conflict. So even though tribal people have few material possessions, they are not poor by their own standards. Lacking experience with serious want and poverty, they assume the world meets human needs.

Indigenous people have difficulties, of course. Traditional herders suffer when animals get sick, or when the winter is too cold or the summer too dry. Interpersonal conflict results when people are sexually jealous or just dislike one another. People can sprain their ankles and be frustrated in the hunt. In short, indigenous people experience the ills that flesh is heir to.

However, they address difficulties differently than we do. Assuming nature is sufficient to meet human needs, which is just the opposite of the scarcity assumption in our culture, they see no reason to increase extractions from the earth. Society does not need to develop newer, better, faster, more powerful tools and techniques. In other words, progress is unnecessary. People can rely upon traditional methods.

This perspective affects the indigenous view of time. Whereas we tend to view time as a straight line, they tend to view it as a circle. In its essential nature, according to typical tribal world views, the world is the scene of endless repetition. Unlike the Myth of Sisyphus in our culture, such repetition is not viewed as meaningless, frustrating drudgery. Instead it is celebrated, as many of us celebrate springtime and the return of birds each year. The endless cycles of days, weeks, moons, years, and lives are good in themselves, apart from any larger cosmic or human drama of progress in linear time, because the world is good as it is.

Sharon Venne, a member of the Cree nation who lectured at the hearing on indigenous concepts of time and place, related a positive attitude toward the present to indifference toward linear time. She said that her ten-year-old son "does not know what day it is to-day, all he knows is, it's a good day to be alive. . . . So I teach my son about time," she continued, "his own time, to rise with the sun and to go to bed with the sun, and to know that the sun is important and to know that the circle of life is important, but everything else is not that important."

Another consequence of viewing the world as good is avoiding invidious comparison with a separate, or future, reality. In our culture's biblical tradition, the world is said to be good as created, but then tainted by human actions. The result is the perception that nature as we know it suffers by comparison with heaven, and with the future kingdom of God on earth. Typical indigenous world views do not value a separate reality, or future time, more than endless cycles of earthly birth, growth, decay, and renewal.

Meaning, Security, and Individualism

How can people find meaning in endless natural cycles? Meaning is a function of connections. Because they are social by nature, people generally find life meaningful when they perceive their activities as contributing to community life. For most people, community life and historical traditions that the community upholds form the larger context that gives meaning to individual life, and particular projects.

The relationship between meaning and connection exists in many areas. In intellectual contexts, it is only when a statement is connected to a group of beliefs that we think a person knows what she is talking about. We would not say that an eight-year-old's recitation of $E = mc^2$ is meaningful to her unless she could connect that formula with other statements related to the theory of relativity. Without that connection, she may mean no more than a parrot reciting the same formula. If the statement is meaningful to us, it is because we can make the appropriate connections. More generally, the question, "What do you mean?" calls most often for an explanation of connections between what was said and other statements, beliefs, or observations.

Meaning in work, too, depends on connections. Consider two people who take money at tollbooths on a toll road. The one who has no interest in transportation and can find no connection to her other interests will probably find the work entirely meaningless. The other will likely see more meaning if she is interested in transportation problems, automobile makes and models, or the way people vary driving techniques when the weather changes. Her work is more meaningful because she connects it to other interests.

Existentialists conceived of individuals as essentially divorced from nature and from one another. They also rejected belief in life after death. They agonized over the meaning of life because meaning comes with connections of the sort they denied. Theirs was an exaggerated representation of the diminished sense of roots in our society.

Many people attempt to gain meaning precisely by finding such roots. Geneological searches and ethnic identifications are among the signs that people find life's meaning in connection, and that connections characteristic of our society are often insufficient.

Our lives are generally fragmented, making us prone to a sense of meaninglessness. The division of labor in society most often divorces the roles of parent and teacher, spouse and worker, colleague and neighbor, and so forth. For example, the demands of competitive excellence and "giving 110 percent" at the office do not take into account requirements of family and home. On the other hand, depictions of family life usually omit realistic representations of the work that adults must perform outside the home to afford family possessions and activities.

The disconnection is evident in television commercials. A commercial for a deodorant features a man (or woman) playing racquetball to virile exhaustion. Psychologically driven to win at work and at play, the man uses deodorant to get a winning edge. In the next commercial a man is relaxed with his child, not driven at all, just hanging around watching with amusement as his child plays with a new toy. Could the same person realistically fill both roles without great internal tension? Impatience with children, high divorce rates, and overconsumption of alcohol and other drugs, suggest not. Roles in tribal societies allow for more consistency of character. Tribal people emphasize connections among people, among various roles people play in their lives, among different stages of life, among generations, and between life and death.

Tribal cultures effect a sense of connection by marking major life events with traditional rites of passage, such as initiation ceremonies. While sometimes involving anxiety and pain, these ceremonies give people a sense of belonging to an ongoing, meaningful order within which they are connected to ancestors, contemporaries, and succeeding generations. Even death can be faced more securely in this context.

Galsan Tchinag, of the Tuwina people in Mongolia, addressed this theme at the hearing. "For the Tuwina", he said:

> there is no such thing as final death. I do not die, I simply change the form of my being. Today, I am a person, tomorrow, I may become a stone and will someday crumble to earth in order to grow up into grass to become a yak, a dog, etc., and finally to become a person again. Life is an eternal cycle, and whoever enters it once cannot be lost from it. So we do not see death as something terrible.

Here the cyclical concept of time is tied not only to appreciation of the earth and life as they are at present, but also to a sense of personal security, even in the face of death, and to beliefs about the relatedness of all beings on earth. Themes of security, appreciation, relatedness, and meaning in life are bound closely together in much indigenous thought.

One result of this sense of security is individuality. It is ironic that American society prides itself on the development of the individual. Insecurity stemming from ambivalence about roles, expectations, and values in our society results not in individuality, but in "the lonely crowd." Consumer products are mass-marketed on the assumption that people are so insecure that they will grasp at any toothpaste, hair style, or shoe that promises, however unrealistically, to provide social acceptance.

Another factor militating against genuine individuality in our society is our reliance on powerful technologies and intricate interdependencies that make deviants particularly dangerous. Rather than promote dangerous eccentricity, we tie individualism to conformity. We say, for example, "You pack your own chute," to indicate each individual's responsibility for his or her own welfare. Because the chute must be packed in just one way in order to work, the individual's responsibility for her own welfare turns out to be the requirement that she do exactly as she has been instructed, just like everyone else. This is individualism without individuality.

By contrast, according to many anthropologists, individuality flourishes in many indigenous societies, where technologies are less dangerous and people are more secure in their sense of self. For example, the Xavante of the Amazon rain forest are as outspoken as heroes in our westerns. The Semai of Malaya prohibit almost all interference in another person's affairs. Paul Radin writes:

> [T]he outstanding positive features of aboriginal civilizations . . . are three: the respect for the individual, irrespective of age or sex; the amazing degree of social and political integration achieved by them; and the existence there of a concept of personal security which transcends all governmental forms and all tribal and group interests and conflicts.

Rootedness and the Expansion of Society

Indigenous people typically integrate a sense of place with their strong sense of meaning and self. They work with local materials to perform almost all tasks in life, ceremonial and practical. Tasks are themselves conceived to make life possible and attractive in the particular locality where they live.

For example, providing housing among the Baka of the rain forest is necessarily different than among the Ju/'hoansi of the desert. They use different materials to make dwellings designed for different purposes and lengths of time. Baka dwellings use leaves to keep out the rain so that people can stay dry inside, whereas traditional Ju/'hoansi dwellings are used more to store belongings than to house people. Socialization is inseparable from rootedness in such societies because people are socialized to perceive and do things that make sense largely in view of the local climate, flora, fauna, and topography.

Laurie Goodman of the Navajo nation made the point this way at the hearing:

> [O]ur homeland is what makes us who we are. It identifies and makes clear our purpose in life. Soon after birth we have a ceremony, where our umbilical cord is buried in our land to symbolically tie us to our land and people forever. . . . The land becomes a part of a person and of our religion. Therefore, when you separate native people from their lands it is equivalent to taking away their will to live. It is hard to convey this concept to a white man who thinks nothing about selling his home and moving on for better economic opportunities. We don't have that choice.

Ian Zabarte of the Western Shoshone Indian nation made the same point at the hearing:

> These lands contain the patterns and lifeways which our people need to carry on our traditions, much like DNA carries the messages of the human body. If these lands are uninhabitably destroyed, we have nowhere to go, no reason to go on. We cannot discover some other land to inhabit. Our culture and tradition dictate that we remain within our traditional homelands as caretakers. This is our responsibility.

Sharon Venne's testimony at the hearing also included the importance of place. She recalled what a Cree elder had said to her. "If you want to sell the land, our indigenous land, then the best thing to do is to take a gun and to shoot your son, because if you sell the land, what kind of future have you left for him and his children? All you've left is dispossession and despair. Save them from that. Kill your son now."

One way that indigenous people express their sense of place is by extending the social sphere to include nonhuman inhabitants and constituents of that place. Esther Yazzie, speaking at the hearing about the importance of the mountains to the Navajo people, said that they identify with its major constituents. "We are the water, we are the fire and we are the air. . . . We are the forests, we are the tree of life, we are the plants, we are the animals, which live on the mountains, from the smallest to the largest."

Similarly, Galsan Tchinag said of the Tuwina people:

> The sky is our father, the earth our mother. We are their children. All people are brothers and sisters, no matter what their language or skin color may be. Moreover, all other living beings are our brothers and sisters. . . . Even the trees, the grasses, the stones are related to us. So we see the larch tree as our brother, the glacial mountains as our ancestors, and the stream as our sister, and every stone we encounter on the various ways through life we see as our brother whom we must treat well and from whom we may expect help.

Students of tribal cultures confirm the testimony of these indigenous people. In many such cultures, especially those of foraging people, the social sphere is extended to include nonhuman constituents in systems of reciprocal (noncommercial) exchange governed by principles of debt and reciprocity.

In Ojibwa stories people engage in reciprocal exchanges with animals that they hunt. Hunters offer animals such items as clothing, earrings, and tobacco, which animals are believed to appreciate. Respect must be shown to animals as well or they will not give themselves to hunters, and the result will be hunting failure. This is similar to the kind of reciprocal exchange and respect that must be shown to human exchange partners.

The Xerente in central Brazil similarly include the peccary, a wild pig whom they hunt, in their idea of the social world. The peccary are assumed to live in their own society, which resembles Xerente society. People can make agreements with the peccary, and when they do, these agreements are generally considered more binding than agreements among human beings.

The Makuna of southeastern Colombia have similar views about animals and fish. If proper offerings are not made, they will die out and no longer be available for food. When fish are spawning they are dancing in a ceremony and must not be eaten because their ceremonial paint is poisonous.

Anthropologist Nurit Bird-David summarizes the views of three other foraging societies, "the Nayaka of South India, the Batek of Malaysia, and the Mbuti of Zaire." She finds that "each group has animistic notions which attribute life and consciousness to natural phenomena, including the forest itself and parts of it such as hilltops, tall trees, and river sources."

What is more, "the natural (humanlike) agencies socialize with the hunter-gatherers. . . ." For example, "the Forest visits the Mbuti camp, plays music, and sings with the people. . . ." Just as sharing food is general among foraging people, they think "the natural agencies give food and gifts to everyone, regardless of specific kinship ties or prior reciprocal obligation." Bird-David maintains that their world "is a cosmic system of sharing which embraces both human-to-human and nature-to-human sharing. The two kinds of sharing are constituents of a cosmic economy of sharing."

The Noncommercial and Sacred

Many indigenous people show respect for nonhumans also by restricting commercial exchange, as we do with human beings. We reject the sale of human beings even when those subject to sale are protected from harm. It is illegal, for example, to sell babies for adoption, even when all safeguards normal to adoption are in place, because buying and selling people seems disrespectful to humanity as such.

Most people in the United States think similarly that because sexual expression is so personal, it should not be bought and sold, and prostitution should be illegal. Many feminists support the legal ban, arguing that prostitution degrades women (the majority of prostitutes) by placing them (their bodies) in the marketplace. Evidence that legalization would reduce disease and physical abuse among (women) prostitutes has changed few feminists' minds.

Indigenous societies typically accord this kind of respect to at least some nonhuman constituents of the environment by restricting exchange. For example, the Weyewa people on the island of Sumba in eastern Indonesia trade extensively, creating debts and social bonds. But they refuse to include their animals, mostly cattle and pigs, in monetary, commercial transactions. Instead, they use these in exchanges that accompany matrimony. Wife-givers offer cloth and pigs to the groom's family, and the groom's family offers cattle and gold.

If these items were all translated into money, then some people could become rich while others remained poor, as in our society. Persistent mutual indebtedness, and the community's moral economy, would be destroyed. People in our society opposed to prostitution and baby-selling also believe that the community's moral economy is undermined when everything people want or need can be bought and sold. This is the theme of the book and movie "Indecent Proposal," which relates the story of a married woman who sleeps with a rich man for $1 million so she can help her husband professionally.

The nomadic Gabra of Kenya's Chalbi desert maintain similar practices regarding camels. Camels are basic to their way of life because they supply needed transportation. Although camels are often lent to others, their sale is strictly prohibited.

Many indigenous people consider the land that they hold as a commons to be sacred, and forbid its individual ownership and exchange. Its sacred nature also precludes many exploitative activities, such as mining. Thomas Banyacya, representing the Hopi nation at the World Uranium Hearing, said this about the land of his people:

[I]t's surrounded by four sacred mountains and the old people knew that that must be left in the natural state. It's the spiritual center of that country. They say that that contains more mineral resources than any other place, but we must leave it in a natural state. We must protect it that way, we must not allow anyone to come and dig the mineral resources up. . . .

Speaking also of land in the southwestern United States, Manuel Pino, representing elders from Acoma and the Laguna Pueblo, objected to mining on Mount Taylor. "[O]ur sacred mountain," he said, "was desecrated."

They stuck the world's deepest uranium mine shaft into our sacred mountain. I ask you, people of Salzburg, how would you feel, if we came here and stuck a jack hammer into Salzburg Cathedral? That's the way we feel about what is being done to our sacred mountain. It is our life, it is our existence, it is our future, it is our present and it is our past.

The same perspective was expressed by Joan Wingfield, an Australian aborigine representing the Kokotha people. She compared mine operators' removal of sacred stones at Roxbury Downs in southern Australia with knocking down Stonehenge. Ian Zabarte, a Western Shoshone, testified also to the sacred nature of the land. "It is our belief," he said, "that all land is sacred." He concluded, "We cannot continue to destroy them. . . . We should take care of our lands and try to clean them up, no matter what it takes, no matter how long."

It is not surprising that indigenous people often, perhaps characteristically, attribute sacredness to the lands they inhabit. The local environment supplies all of the people's needs. Lacking belief in a transcendent reality superior to earth, this land functions analogously to God in Western religion. It is the reality within which people live, move, and have their being.

Another respect in which the land for foragers is analogous to God in Western religion concerns parentage. Just as in Western religion people view themselves as children of God, the foraging people discussed by Bird-David:

> . . . . regard themselves as 'children of' the forest. . . . For example, . . . the Mbuti often refer to the forest as 'father' and 'mother' . . . but also . . . describe it as the source of all spiritual matter and power, including the vital essence of people's lives. . . . In a similar vein, . . . the Nayaka not only refer to natural agencies (especially hilltops and large rock formations) by the terms *dod appa* (big father) and *dod awa* (big mother) and to themselves correspondingly by the terms *maga(n)* and *maga(l)* (male and female children) but also say that dead Nayaka become one with the forest spirits.

Such people would no more sell land for commercial advantage than, to paraphrase W. C. Fields, stick a fork in their own mother's back.

Environmental despoilation was equated repeatedly at the hearing with violations of Mother Earth. Guy White Thunder's perspective on the Black Hills in South Dakota is representative:

> The Indians call the Black Hills our mother's purse. That's Mother Earth's purse, where there's a lot of minerals, gold and uranium—all these things that Mother Earth had in her purse. And a big corporation comes over there and they know about this gold and they steal it from our mother, they steal from the Mother Earth, and the Indian people do not like this at all. They should leave the ground the way it is. It's pretty without people cutting it up. One lady says, what would you look like if I cut up your face and your body?

Indigenous World Views Are Nature-Friendly

The last chapter showed that some indigenous people achieve peace and prosperity without oppressing human beings. The present chapter, summarized below, shows that subduing nature is equally unusual for these people.

Native people lack our motivation to seek increased power over nature. We seek such power largely to extract more of what we need from nature, in light of our belief that nature provides insufficiently for us. Tribal people, especially foragers, typically view nature as supplying abundantly, so they are not motivated to seek the power that we have attained. They would not destroy natural patterns to prepare for a new, better world, as in the Bible's Book of Revelations or Marx's book of revolutions. They generally try to maintain earth's natural systems.

Tribal personal security, individuality, and sense of meaning also foster environmentally sound practices. In our culture, people often seek favored social positions, individual distinction, and a sense of meaning by producing or obtaining material goods. This is captured in the saying: "Who dies with the most toys wins." Feelings of self-worth are thus tied to increased extractions from the earth. Indigenous people are less inclined toward such extractions because they provide differently for personal security, individual distinction, and a sense of meaning.

Indigenous societies tend to be nature-friendly because their populations are usually stable. Increasing human populations aggravate environmental degradation caused by human beings.

Perhaps most significant is the social representation of nature in indigenous thought that results in restrictions on commercial exchange, analogous to those we make on commercial exploitation of sex and procreation. Such restrictions among indigenous people limit the kinds of commercial exploitation that often prove environmentally destructive. If trees cannot be owned individually and sold, they cannot be cut down to export lumber. If camels cannot be traded for cash, no one has an incentive to own so many as to exceed the carrying capacity of the commons.

Another result of the social representation of nature leads more directly to environmentally responsible behavior. Plants and animals are represented as engaging in reciprocal exchange with people, just as people do with one another. It is essential that such exchange be balanced, that neither party take advantage of the other. This disinclines indigenous people to exploit resources to the maximum. As with human exchange partners, they limit exchange demands in order to maintain the relationship long-term.

This perspective resembles an aspect of Aldo Leopold's Land

Ethic. Leopold would have people be plain members and citizens of their biotic communities. This is a political metaphor for the tribal injunction to respect all constituents of the environment as we respect human relatives and exchange partners.

Finally, tribal people often view nature, or many of its significant features, as sacred. It occupies the place of God the father in Western religions. Just as we limit exploitation of our human parents, and consider respect and gratitude more appropriate than exploitation in relating to the divine, indigenous sacralization of nature inclines tribal people to avoid environmentally destructive practices. If the mountain is sacred, uranium cannot be mined there. Worldwide, find indigenous people and you will find relatively undisturbed ecosystems with abundant biodiversity; find relatively undisturbed ecosystems and you will find indigenous people. The overlap is striking.

In sum, indigenous people largely conform their behavior to something like Aldo Leopold's Land Ethic, which values nature for itself. This value structure among indigenous people is tied closely in *some* indigenous societies to practices that are, relative to our own, friendly toward people as well as toward the environment.

The upshot is that the best way to help people is to adopt a value structure similar to that of these native people. If we really care about human beings, the most effective way to express that care is to broaden our conception of what has value for itself to include soils, water, lakes, mountains, plants, animals, and the earth itself. I do not suggest that we *pretend* to value these things, but that we *really* value them, as only in this way can we approximate the views of indigenous people.

The implication is that we should reduce attempts to manipulate nature to human advantage. Possibilities for doing this in *our society* are discussed next.

Chapter 9

Implications

Promoting Change

Friday evening, after a week of testimony, the hearing ended with a tree-planting ceremony high on a hill near Salzburg. The tree's growth would represent growth in awareness and earth-centered values.

I wondered if awareness and values would grow as needed. If I am correct, people need to change from exploiting nature to valuing it for itself, from increasing human power over nature to reducing it, and from seeking a growing economy and more consumer goods to preferring a reduced volume of commercial exchange. Can ideas that fly in the face of what most people consider universal common sense be adopted in time to avert additional human oppression and environmental breakdown? If not, the future seems bleak.

At this point in some futurists' work the author points out that, bleak as prospects may appear, there are positive tendencies already operative in society that are likely to save us after all. The rest of the book is evidence for, and glowing descriptions of, these tendencies. No one has to change course. Progress is reaffirmed, the reader is pleased, and book sales are brisk.

Another approach is to find that other people cause major problems, so *they* have to change. Welfare mothers cause budget deficits, "special interests" block needed legislation, criminals make streets unsafe, and so forth. Like the elimination of Jews from Germany, severe measures are justified because you cannot make omelettes without breaking eggs. (People who say this identify with cooks, not eggs.)

An environmental version, so-called lifeboat ethics, identifies the major environmental problem as overpopulation, especially in the Third World. To save the earth we must allow millions of poor people to starve while there is plenty of food for everyone. Never mind that our culture (for commercial reasons) weakened their traditionally effective methods of population control and ruined traditional methods of obtaining food.

This is just another case, characteristic of commercial-industrial thinking, of responding to jeopardy with proposals for human oppression. The proposal is inherently commercial, not environmental, because only poor people in such places as Sudan and Bangladesh are required to starve, not the more densely populated Japanese, who can pay for food.

Rather than blame others, I believe that to solve our problems, *we* must change. We can see this by relating changes needed to values and beliefs already held. Chapters One to Six, for example, show that projects to master nature in the human interest cause much human oppression. People opposed to human oppression have reason to reject our anthropocentric approach to nature in favor of indigenous approaches. Contrasting our culture with (admittedly imperfect) indigenous cultures in which there is less power concentration and in which nature is revered for itself, we conclude that we have in some important sense taken a wrong turn *as judged by our own revulsion to inhumanity.*

Family Values

Family values and friendship are also at odds with commercial-industrial exploitation of the earth. I am struck when listening to the news by the weakening of strong family ties, and associated problems of loneliness, mental health, drug addiction, and crime. I think our commercial-industrial culture weakens families, whether they be nuclear families, traditional extended families, avant-garde communes, or homosexual unions. Generally speaking, friends and family provide not only what money cannot buy (such as feelings of warmth, security, and belonging), but also what money can buy. Friends are often available to talk over a problem or look after children when parents are unable. Family members traditionally cook meals, rear children, care for old people, and clean house in cooperative, noncommercial arrangements. Yet these are all services that can be commercialized. Paid counselors or therapists can discuss personal problems, commercial day-care services can care for children, and special retirement communities can house old people, while restaurant chefs do the cooking.

Replacing friends and family with paid workers adds to the gross domestic product (GDP—the monetary worth of goods and services produced in the country), and so contributes to economic growth. Increasingly, both parents are expected to work outside the home, even when children are small. When new acquaintances ask, "What do you do?" and a woman answers "I stay home with the children," her right to participate in adult conversation is jeopardized. A man giving that answer evokes pity or disgust. Such reactions motivate commercial purchase of child-care services.

Other motivating factors are advertisements that equate the good life, even love of children, with something that money can buy. Really caring parents work outside the home to afford expensive toys, clothing, computers, and lessons of all sorts for their children. This increases GDP through purchases of consumer items and child care. In sum, taking care of one's own children is discouraged because it conflicts with commercial interests.

Social norms also discourage old people from living with younger relatives. Old people are encouraged to think they would be a burden, and that self-respect requires independence. This promotes economic growth through establishment of retirement communities, and purchases of child-care that might be free in extended families.

Extended families and the free services they provide are discouraged also by the expectation that people should be willing to relocate to further their careers. Far away from parents, siblings, neighbors, and friends, people increase GDP when they buy from strangers services that friends or extended family members would have provided free. Worker mobility increases GDP also because technological change can alter the location of productive work. If coal is not needed at the moment, coalminers should move to where clothes need sewing, computers need building, or hamburgers need tossing.

The minimum wage must be kept low, it is often argued in the United States, because a higher minimum wage would increase unemployment. Businesses willing to pay $5 an hour to unskilled workers would not pay $8 an hour. Low labor costs help commercial services, such as fast-food restaurants, compete with free services within the family.

The general pattern resembles enclosure movements. Before enclosures, people typically live in local, mutual dependence, with minimal commercial (monetary, exchange-value) transactions. Other people, with exchange-value interests, figure out how to make money from resources formerly available free. They claim that efficient use of resources helps people win the battle against endemic scarcity. So peasants are kicked off their land (to grow sheep for wool), rain forests are bulldozed (to raise cattle for beef), and those who had been self-sufficient in noncommercial societies are rendered incapable of living decently (or living at all), without selling their labor.

Similarly, the commons area that we call home is impaired when endemic scarcity is combatted by more people simultaneously working elsewhere and buying what had been free at home. When the home satisfies fewer material needs, people reduce time commitments and emotional investments. In the end, the home can no longer supply enough of what money *cannot* buy: security, warmth, and belonging.

There is no conspiracy in corporate headquarters to destroy family life. It is simply the logic of the situation. Independent, lonely, atomic individuals are generally better customers than people in close families and communities. So associating human well-being with a growing economy fosters ideas and practices that tend to destroy family life.

Crime, Pornography, Drug Abuse, and the Work Ethic

Often related to the breakdown of families are increases in crime, pornography, and drug abuse, and erosion of the work ethic—the willingness to work hard, defer gratification, take pride in work, and accept responsibility.

Industrialization has made production increasingly easy, so a major impediment to economic growth in wealthy societies is a lack of demand. This is why advertising is needed to stimulate the demand for goods and services.

Advertising often stimulates demand by claiming products reduce the work needed to attain whatever is wanted, whether it be a good-looking lawn, clean laundry, a larger vocabulary, or happy, healthy, well-behaved children. An underlying message is that leisure and play are desirable, but not work. Ads also portray people as impatient: they want everything five minutes ago, and they don't want to work for what they get.

These messages are all contrary to the work ethic, according to which solutions to problems come slowly and through hard work. Pride in the result is sufficient compensation for pain in the process, which is considered character-building.

The relationship to crime and drug abuse is not hard to see. Stealing and taking drugs are often shorter and (in the short-run) easier routes to personal gratification and happiness than the alternatives of hard work and acquisition with hard-earned money. No wonder crime rates and drug abuse continue to rise in the United States despite strong and costly law-enforcement efforts.

The relationship of drug abuse to commercialism is evident in school candy sales, often used in the United States to support a school activity, such as a sports team or choir. My observations as a parent suggest that candy functions for children much as drugs function for others. Like drugs, candy provides instant gratification; children like the taste. Also like drugs, candy often has bad aftereffects, although in the case of candy it is usually only tooth decay. More important, like drugs, candy tends to replace more wholesome choices, such as more nutritious food, and so detracts from proper nutrition. Finally, this replacement tends to have long-term consequences because instant gratification can be habit-forming. Children used to candy are more reluctant than others to experiment with nutritious foods. The result can be a less healthy life-style, more frequent illness, and reduced development of the individual's innate capacities.

Candy is certainly much less potent than controlled substances, but its function for children is structurally identical to the function of controlled substances for others. It is to drugs as school elections are to national campaigns. School candy sales recruit children as both customers and "pushers." Advocates claim that these sales not only raise needed cash for the school, but teach children how the free market works. I could not agree more.

I was surprised when visiting Salzburg to find that one of our hosts, Christa Jecel, shared my perspective. Her medical practice includes caring for children at local elementary schools, where children can buy candy from machines. When she recommended removing the machines, she was told that this could put the man who owns them out of work. School officials were unconcerned by their doctor's warning that his work prepares children for drug abuse.

Attempts to increase consumer demand invite pornography as well. People who have enough money to be targets of advertising often own all that they really need. Some method of getting their attention is required before they can be convinced to buy more. Sex is used in ads for this purpose. Beautiful women in bathing suits surround a new car; beautiful twins appear in ads for chewing gum; the man wearing appropriate jeans, or drinking the right beer, heads off with a woman who seems to be in heat, and so forth. Just as catnip is supposed to be so attractive to cats that they will harm themselves by eating too much, sex appeal makes products so attractive that people buy more than is good for them.

People become habituated to increasingly explicit visual depictions of sex, as clothing advertisements approach soft porn. The sexually enticing poses in Victoria's Secret catalogs are taken for granted, as if there was no hint of pornography. But in 1995, when Calvin Klein used underage models to similar effect, the uproar over charges of kiddy porn prompted alteration of the ad campaign.

Creating Jeopardy Is Good Business

As already noted, commercial-industrial culture fails in an area often considered most important—individual satisfaction, development, and fulfillment. Worried, dissatisfied people are generally good customers.

Many worries that stimulate consumer demand are well justified, as they stem from social evolution, not just marketing hype. Where rich and poor live in close proximity, and the poor are increasingly desperate, improved home security systems make sense. If economic growth is equivalent to improvements in human well-being, then a public terrorized into buying such systems is better off than one living in relative safety because economic activity is greater.

The creation of many other problems also can stimulate economic growth. The Alaskan oil spill of 1989 stimulated $2 billion in cleanup expenditures and was therefore good for the economy. Air pollution in the United States causes illness that requires $40 billion in medical expenditures annually. These are included in the GDP we are trying to maximize. By this perverse logic we are better off with illness than with health.

Of course, economic growth can be stimulated also by expenditures incurred to avoid environmental and health problems. Building a new fleet of double-hulled oil tankers would stimulate economic growth, as would more labor-intensive agricultural methods that reduce the use of health-harming herbicides and pesticides. But GDP does not distinguish between prudent measures that avoid or lessen problems and foolish measures that create problems requiring expensive cleanup, making it an unrealiable guide to human well-being. However, increasing GDP is crucial to our culture's commercial-industrial thinking and the polestar of those seeking to eliminate barriers to international free trade.

In sum, people in our culture have several reasons to consider changing course. The polestar of the current course is an unreliable guide to human well-being. Its pursuit fosters inhumanity; the breakdown of families; dissatisfaction; increases in crime, pornography, and drug abuse; and decreases in personal responsibility.

Rejecting Utopian Thinking

People who criticize their societies often outline a utopian ideal to serve as a goal. Plato's *Republic* is perhaps the most famous. I do not specify a utopia for us to realize.

I find arrogant the claim that we know how to produce a concrete, desired result in human affairs. Consider the provision of running water in rural Mexican houses, which was supposed to benefit women, whose job it had been to get water from the village well. Wife abuse increased significantly. Social pressure was no longer applied to offending husbands because bruises were no longer exposed to public view at the well. Fortunately, in this case, the government subsequently set up communal grain grinding that served, incidentally, to expose spousal abuse, which then diminished. The point is that people cannot obtain predetermined results in the manipulation of society because there are too many unknown interactions.

(In the United States air-conditioning now isolates people in their own houses in hot weather, when formerly neighbors met outside while trying to keep cool. Could reduction in neighborly interaction be promoting (supposed) increases in domestic violence, making air-conditioning a feminist issue?)

Skepticism about our ability to obtain predetermined results
through manipulation of society (or nature) discourages my elab-
oration of a utopian ideal, because I recognize that no matter how
worthy the ideal seems, I do not know how to attain it. More fundamentally, I avoid elaborating such an ideal because I am uncer-
tain what the best society would be like. I am not sure what ideal
would be worthy.

I have drawn inspiration from some indigenous societies, but I realize that our society cannot duplicate them. We cannot in any foreseeable future abolish nation-states, eliminate divisions of labor,
avoid the use of electricity, and refrain from commercial exhange.

Governments of nation-states will be needed in the foreseeable
future for most of their current functions. They also will be needed
to cushion the effects on many people of required changes. If uranium mining ceases, uranium miners will need help redirecting
their lives. Nation-states will serve additionally by entering into,
and monitoring compliance with, worldwide agreements required
to address such problems as ozone depletion and global warming.
Analogous considerations show the continuing need for some division of labor, electricity use, commercial exchange, and sophisticated birth control (to avoid the infanticide practiced by many indigenous people). So whatever I have learned from indigenous
people must be used in a society different from theirs. I cannot en-
vision this new society in any detail.

Aspects of our current culture needed in the foreseeable future
may be refractory to admirable aspects of indigenous cultures. Having the best of both worlds may be simply impossible, and any
utopian ideal of this genre completely unrealistic. We simply cannot know in advance of experimentation.

Invention Is the Mother of Necessity

Rather than describe an ideal society, I propose a direction of movement. We should reduce our (human) impact on the earth's ecosystems, becoming increasingly, as Aldo Leopold wrote, "plain mem-

bers and citizens" of our biotic communities and of the biosphere as a whole. The present book has been devoted primarily to explaining why it is desirable *from the perspective of our current values* to change from increasing to decreasing human dominance of nature. We need to change direction in order to reduce the oppression, dispossession, murder, and starvation of people, as well as to resuscitate community and family ties; stem the tide of crime, pornography, and drug abuse; and promote genuine human fulfillment. Experience to date justifies replacing our secular faith in human benefits stemming from power over nature with faith that reducing power over nature serves people better.

People in our culture often say that necessity is the mother of invention. This flatters our inventive culture. We believe that because inventions meet preexisting needs, they are making the world better, at least for people. But in many cases the relationship is reversed. Invention is the mother of necessity.

Consider genetic engineering. Genetically engineered human growth hormone, which helps people become taller, is already given to people likely to be extremely short due to genetic or other abnormalities. The producer, Genetech, recommends the drug be made available to anyone whose height is likely to fall in the lowest 3 percent, but there will always be a lowest 3 percent (except where all the children are above average). Because height is generally advantageous in our society, many medical researchers wanted the new hormone for their own children of normal height.

These reactions suggest that, when commercially available, this therapy will catalyze: diminished self-sufficiency (an increasing percentage of people will not be able to grow without prescription drugs); increasing social inequality (poor people will not be able to afford the drug); and possibly long-term negative health effects (for all we know recipients may be at increased risk of heart disease or cancer in middle age). This is the kind of case where need for the new faith (that efforts to control nature in the human interest will backfire) should be clearest. But most people concerned with promoting the human good do not see it.

The greater the success of genetic engineering, the greater the peril for humanity in general. Consider the consequences of genetically engineering increases in human intelligence. "Intelligence" will probably be equated with the ability to invent more ways to manipulate nature in the supposed human interest. Nations will compete to refine and distribute intelligence-enhancement techniques. Parents unwilling to have their children treated will be accused of violating their children's right to succeed, and many will go into debt to afford treatments, as they now do for their children's college education. An intelligence gap will be created between rich and poor, justifying Social-Darwinist claims.

Invention is the mother of necessity in these, and many other, cases, due to competition. Whatever invention improves performance becomes genuinely needed by all competing parties. Because height gives competitive advantage, the use by some people of human growth hormone creates a genuine need in others. The invention of cheaper manufacturing processes, new weapons, and crops with greater yields or more resistance to drought or pests all create the need for others to follow suit.

Only occasionally is the harm of the invention taken seriously enough that all parties agree to ban it, as with steroids in athletic competition. Most often, competition is so intense that the invention is introduced by those seeking short-term competitive advantage without regard to long-term negative consequences. Cheaper manufacturing processes may be dangerous or lead to psychological stress, new weapons may precipitate more devastating wars, and crops with geater yields may require expensive fertilizers that poor farmers cannot long afford, leading eventually to the widespread starvation that followed the "Green Revolution."

Invention is the mother of necessity also because the physical infrastructure of society can be altered by inventions such as the automobile. Places where people need to go are moved farther apart, with little or no provision of public transportation. People with no interest in either automobiles or competition must nevertheless have cars in order to perform ordinary life tasks in the altered physical environment.

Some inventions are beneficial, of course. But so long as we believe that the world as we find it is insufficient to meet human needs, we will be insufficiently critical of inventions. Thinking that we need a host of inventions to combat endemic scarcity, we will continue to accept inventions that create more needs than they meet. A necessary means of avoiding this result is a new value structure that celebrates nature as we find it.

New Faith and Values

We must come to respect, value, and revere nature for itself, and as it is. Anthropocentric reasoning has so far been ineffective in our culture. New faith (in the backfire effect of manipulating nature) must be combined with genuine respect or reverence for nature as it is. Only when people (in our culture) revere and value nature for itself will they refrain from manipulations that imperil humanity.

Adopting nonanthropocentric values that are common among indigenous people does not require rejecting monotheistic religions. The Vatican's representative to the World Uranium Hearing, Archbishop Emanuel Milingo from Sambia, helped me to see how reverence for the earth can be combined with worship of God. The following is taken from the prayer he offered before the start of Wednesday's testimony.

> Oh Father God, we thank you for giving us this occasion of being together. . . . I will offer to you praise and glory and all our activities today. . . . Oh Mother Earth, for the offences that you manage instead of appreciating the life-giving force which you distribute through all that we eat, through all that we breathe, all that makes us move. We ask you to pardon the abuses you have received instead of gratitude.

If Roman Catholics can revere the earth without abandoning Christianity, other Christians, as Jews, and Muslims can revere nature without abandoning God.

This coheres with religious opposition to genetic engineering. Some Christian groups, including the Catholic Church, see human genetics as a gift from God to be accepted, not manipulated. Biblically based religions could be cornerstones of the new faith and value system because, as Archbishop Milingo's prayer indicates, the Bible is compatible with nature-centered views.

The Bible's flexibility stems in part from the common practice of taking a single passage or verse to justify major religious or social practices. Jews have built many laws of kosher eating on a verse that prohibits cooking a kid in its mother's milk. Jehovah's Witnesses have forbidden members to receive medical blood transfusions because of their interpretation of the Bible's prohibition of drinking blood. The Roman Catholic Church takes the command to be fruitful and multiply to imply the impermissibility of contraception.

There is nothing illegitimate, much less insincere, in this practice of placing great importance on limited sections of the Bible. The Scriptures are simply too long and complex for people to give equal attention to every verse. Also, if the Bible is to guide people over the centuries, it must be susceptible to varying interpretations adapted to changing circumstances. So apart from the merits of the particular biblical interpretations given as examples, the interpretative practice they exemplify is unproblematic.

People inspired by the Bible may find warrant to hear the grass scream in the verse, "Holy, holy, holy is the Lord of hosts, the whole earth is full of His glory," on God's postcreation pronouncement that the world is good, or on the appreciation of nature in the Sermon on the Mount. These and other passages could justify revering nature as it is.

This is not Judaism or Christianity as most commonly understood, but interpretations change over time. At one time it was common to find biblical support for racism. Rejection of racial superiority has not detracted from biblically based religions; neither should rejection of species superiority. Some Christian environmentalists, such as Father Thomas Berry, are approaching this view already, so religious people in our culture can modify, rather than reject, their biblically based religions, and agree with indigenous people's social conception of human relations with nonhuman nature.

Resistance to such reinterpretation may stem from the Old Testament's clear rejection of animistic paganism, which reverence for nature resembles, but the Bible rejects ancient paganism largely to repudiate pagan human sacrifice. Today, most human sacrifice stems from the worship of technological manipulations of nature in the (supposed) human interest. I turn next to practical suggestions for moving toward a better society.

Chapter 10

Practical Suggestions

An Alternative Politics

The two most influential political parties in the United States agree that human welfare is paramount and that economic growth helps humanity. They propose to stimulate economic growth with international trade, while fighting crime with more prisons, terrorism with reduced civil liberties, poverty with welfare reform, and declining family values with sermons and tax breaks. Gore Vidal notes that this country has only one political party and it has two right wings.

If I am correct, current approaches will not help because they fail to address the fundamental problem—the incompatibility of a growing economy with many of our other values. To serve these other values, we must find ways to live meaningful, fulfilled lives with reduced reliance on monetary exchanges, which is to say, with greater reliance on ourselves and people we know personally in our families and communities. This chapter contains the outlines of an alternative politics—suggestions for first steps in the right direction. When these or similar steps have been taken, people can assess their effects and decide what to do next.

These steps may seem impractical simply because they are contrary to prevailing political and social movements that promote

economic growth and celebrate scientific and technological progress as saviors of humanity. In this situation, as in most human affairs, practicality is determined by social, not physical, reality. In a
society where prestige is sought by burning more blankets than
anyone else at an annual festival, using blankets to keep poor people warm is impractical. What could serve as motivation? If the society also values generosity and consideration for poor people, an
argument could be made featuring these values. Still, social customs are hard to change, even when they run counter to some of
the society's own values. So reformers should expect change to take
time.

The proposals made here are like this. They appeal to some of
our society's values, but seem impractical because they are contrary
to established customs. In much the same way, abolishing slavery
was considered impractical in a society declaring "all men are created equal." Civil rights, workers' rights, and women's rights all
seemed impractical for generations even though they appealed to
common goals and values. It takes time, but objections about the
practicality of proposals made here can be met by showing that
popular goals and values can be reached and served only with a
smaller economy and less power over nature.

The goals of proposals for a smaller economy include making people more self-reliant, and reliant upon informal, personal arrangements, rather than on formal, impersonal institutions; reducing human oppression; reducing the gap between rich and poor; reducing
the importance of remaining differences in wealth; reducing the human population; and reducing the human impact on the nonhuman
environment. The values served include family closeness, self-reliance, freedom, efficiency, kindness, and personal satisfaction.

Some suggestions concern international affairs, whereas others
concern personal life—things people can do by themselves for
themselves. But I begin with suggested government policies for the
United States. Such policies are crucial in part because our personal
choices are strongly influenced by social context, and national government policies greatly affect society. Also, reduced economic ac-

tivity means reduced job opportunities. To avoid cruelty, the government must cushion the transition to greater self-reliance.

The following policy proposals are meant merely to illustrate the kinds of initiatives possible for reducing reliance on exchange values. I give general outlines of policies concerning agriculture, international trade, and transportation, and shorter sketches of policies on energy, income distribution, and population reduction.

In general, kindness requires gradualism. Ramsey Clark, former Attorney General of the United States and long-time peace activist, pointed out at the hearing that turning off all French nuclear power plants before alternatives are in place would literally kill people. Some of those living in high-rise apartments would die of heart attacks climbing stairs, and others who rely on electricity for heat would suffer hypothermia.

Society's various aspects are so interrelated that changing any one thing quickly and radically throws many others out of order, creating widespread, uncontrollable damage. Pushing through quick, radical change is just another way of acting without much respect for others. The global economy and power over nature must be reduced gradually and with care, like defusing a bomb.

Agriculture

American agriculture is often said to be the most efficient in the world, but it is also the most inefficient. How can this be?

"Efficiency" concerns the relationship between needed inputs and desired outputs. When, all other things being equal, needed inputs decrease, or desired outputs increase, efficiency improves. When, for example, gasoline is used to power an automobile and miles travelled are the desired output, efficiency improves as the car travels more miles per gallon.

American agriculture is the most or least efficient in the world depending on which inputs and outputs are considered. It is the most efficient if the input is person-hours of work devoted directly to raising food. Only 2 percent of the population grows the food that we all eat. In financial terms, however, our agriculture is only moderately efficient. Here the inputs are investment capital and the outputs income. The industry claims to need significant financial assistance from the government. Subsidies are currently about $17 billion a year.

American agriculture is the least efficient the world has ever known when energy is the input and food calories the output. Traditional farmers fed themselves and their livestock from what they grew, so to keep the system going the caloric value of crops had to exceed that expended by people and animals working the farm. Modern American agriculture, by contrast, relies on enormous inputs of fossil fuels in fertilizers and farm machinery. It consumes ten calories for every one it produces.

I suggest an efficiency measure for agriculture different from any of those considered so far. The main reason for agriculture is to grow nutritious food for human consumption, so that is the appropriate output. The inputs are arable land and fresh water, as these are in limited supply and crucial for food production. Their efficient use is needed not only to ensure the human food supply for the indefinite future, but also to reduce human impacts on non-human life forms who rely on the same resources.

Government involvement to improve agricultural efficiency can be justified in part on grounds of national security. We spend billions to secure oil supplies. Food is more important, yet current agricultural practices contaminate groundwater with chemical herbicides and pesticides and deplete needed topsoil through exposure to erosion. Such contamination and loss jeopardize future food supplies.

The government should promote sustainable agriculture, which generally uses smaller machinery, less fossil fuel, little chemical fertilizer, and little or no herbicides and pesticides. Soil erosion is impeded with hedges and trees on the sides of generally smaller fields. Hedges and trees also can host birds and insects central to integrated pest management, which reduces dramatically the need for pesticides. Soils are built in part by rotating crops so that every third year a legume, such as alfalfa, is planted. It adds nitrogen to the soil when plowed under rather than harvested. Crop rotation is used not only to build soils, but also to foil many pests, as those from last season may not eat what is growing on the field this season.

Under current conditions, the transition to sustainable agriculture is not within the means, nor in the interest, of most American farmers. Building soils with crop rotation that includes a legume that is plowed under reduces the amount of land used for cash crops. Soils deprived of chemical fertilizers are less productive for three to five years because they have been damaged by harsh agricultural chemicals. Again, income from crop production is reduced, at least in the short run. Farmers in significant debt must operate in the short run. Planting trees and hedge rows and beginning integrated pest management take time and money. Finally, many farmers do not own most of the land that they farm, so they have no incentive to lose money in the short run, even if they could afford to do so, to promote long-term soil health and productivity.

I suggest that the government subsidize the transition to sustainable agriculture. First, the government should hire experts to teach appropriate techniques through the already existing agricultural extension system. Second, low-interest loans and outright grants should be made available to help farmers make the transition to sustainable agriculture. Third, the government could agree to buy for the armed forces, the Women Infants and Children Program (WIC), and so forth, food grown sustainably. They should guarantee purchase at whatever price covers the costs of production and allows a reasonable profit. Demand should increase supply, according to classical economics, and practice should improve efficiency and lower costs.

The government also should help people purchase the land they are farming, and help them to pass that land along to family members in the next generation. This gives farmers a stake in the land's long-term health. Loans and grants are again in order, as are modifications of inheritance laws to help people keep farms in their families.

Government money should encourage use of land to grow what *people* eat, rather than what cattle and hogs eat. Such animals use about 90 percent of the food energy they receive to maintain their own metabolisms. Feeding ourselves by eating them is an inefficient use of land that could be used to grow food for direct human consumption. Using more land than necessary, we deprive many nonagricultural species of habitat.

Money for needed programs to promote sustainable agriculture can come initially from phasing out currently destructive farm programs of greatest benefit to the wealthiest farmers. Among other things, the government should stop subsidizing the inefficient meat industry. For example, American cattlemen pay only .01¢ per hamburger for use of government land to graze cattle.

If measures are taken to establish sustainable agriculture, agriculture will be more efficient by most standards, but not by that of person-hours needed to produce food. Agriculture will require more people and critics claim that this makes environmental proposals economically impractical. Yet we continue to build the multibillion-dollar *Seawolf* submarine not so much for national security as to keep people working. The major argument for continued destruction of old growth forests is to save logging jobs. It seems we have plenty of people for needed work, but require government help redirecting efforts toward productive pursuits, such as sustainable agriculture.

Because more time and effort is used to produce food, the cost of food will rise relative to the consumer price index (CPI). Meat in particular will become more expensive as government subsidies are phased out. Hardships on poorer people can be avoided in part by changed consumption patterns. People can save money on food by eating less meat, and by preparing fresh food for themselves instead of eating at fast-food restaurants. Significant health benefits and lowered costs of health care would result, freeing some tax money to further help the poorest Americans with direct food subsidies, as through food stamp and WIC programs.

International Trade

Because food grown sustainably is more expensive than food grown without regard for the environment or future generations, American farmers would sell less food abroad and would need protection from the competition of food grown unsustainably in other countries. We do not want to have foreign competition ruin American agriculture, making us dependent on imported food. The best way to protect our farmers from unfair price competition is through import taxes on food grown unsustainably abroad. The tax would make sustainable agriculture attractive to potential exporters worldwide. Revenues from the tax could further the transition to such agriculture at home.

The majority of people in most countries exporting food to the United States would not be harmed by reduced international trade. As noted earlier, in many poor countries the best land, which could be used to feed everyone adequately, is devoted to cash crops for export. This benefits local elites and multinational corporations, but causes widespread malnutrition and starvation. Brazil, for example, "has the worst rich-poor gap in the world."

> Brazilian per capita production of basic foodstuffs (rice, black beans, manioc and potatoes) fell 13 percent from 1977 to 1994. Per capita output of exportable foodstuffs (soybeans, oranges, cotton, peanuts and tobacco) jumped 15 percent. Today some 50 percent of Brazil suffers malnutrition. Yet one leading Brazilian agronomist still calls export promotion "a matter of national survival." In the global village, a nation survives by starving its people.

Reducing agricultural exports could encourage local agriculture for local consumption. Almost every country where there is widespread hunger can be agriculturally self-sufficient.

Regional and worldwide trade accords, such as NAFTA, the European Community (EC), and GATT, are major impediments to improved protection for poor people and the environment. For example, the European Parliament proposed to restrict imports into

EC member countries of logs grown or harvested in environmentally harmful ways. Preliminary indications are that GATT disallows such restrictions, as it may similar measures supporting sustainable agriculture in the United States.

In 1988 the European Court of Justice disallowed a Danish law restricting "the number of different bottle sizes and shapes of both domestic and imported bottles." The Danes were attempting to promote bottle reuse. Reuse is more efficient than recycling because reuse requires only cleaning the bottles, whereas recycling requires melting them down and manufacturing new ones. The Danish law was declared anticompetitive. In 1992 the European Commission similarly disallowed a Swedish ban on importation of "products containing or produced with CFCs, HCFCs and other ozone destroyers." Although all parties agreed that abandoning these chemicals is urgent, the EC enforced the slower pace of other countries.

International trade accords tend to favor large commercial interests, with little regard for environmentally sensitive commerce. For example, NAFTA makes "government subsidies for oil and gas mega-projects . . . explicitly immune to trade challenge or countervailing duties." However, U.S. laws requiring the use of recycled fiber in newsprint may be challenged, and "British Columbia had to abandon reforestation programmes when it was challenged by the U.S. forest industry as an unfair subsidy, and hence contrary to free trade."

GATT threatens many poor people in the Third World with starvation due to application of "intellectual property rights." When a multinational agribusiness improves the seed of a staple crop, farmers will have to pay royalties each time they use the seed. They will no longer be able simply to gather and use seed from the previous year's crop. U.S. seed companies expect to "gain $61 billion a year from Third World royalties." Those who cannot pay will not be able to grow food after seed companies have gained patents on all available varieties. Again, international trade impoverishes the poor to benefit the rich.

The United States should limit its commitment to free trade and to current international agreements. Protecting the poor and the earth have traditionally been the province of national governments in the industrial era. International trade agreements weaken the ability of national governments to do this, and supply no international substitute. So the United States should act unilaterally. We should use tariffs to protect environmental and worker safety standards in agriculture and industry. We should similarly protect systems designed to reuse, recycle, and conserve resources. Some additional ideas are restrict exports of scarce resources that are not managed sustainably; outlaw exports of pesticides and other products considered too dangerous for domestic use; and restrict marketing of unnecessary goods, such as cigarettes.

The above agricultural and trade policies support reduced trade (internationally) and diminished consumerism (domestically, for example, at fast-food restaurants). Resulting reductions in commerce promote efficiency, national defense, long-term sustainability, family farms, human health, the ability of Third World people to feed themselves, and the preservation of habitat for other species.

Transportation

Although implementing the above proposals would create jobs in food production, the overall result of reduced commerce is job loss, for example, in fast foods and in food import and export. In addition, consumers would have fewer inexpensive food and other items produced abroad. However, this does not mean that people's lives would be worse. The following consideration of transportation illustrates how people can live well in a smaller economy with less gainful (monetary) employment.

Governments indirectly subsidize automobile use in the United States. General tax revenues are used to build most state, county, and municipal roads. They pay also for state highway patrols, traffic courts, and national defense strategies (including the Gulf War) used to secure petroleum supplies. Automotive exhaust fumes cause much pulmonary and coronary disease, so people pay indirectly for auto use through health care. If these and other indirect costs were built into the price of gasoline, car use would decline.

Further decline would result if governments were to subsidize public transportation instead. Subsidized the way automobiles are at present, municipal rapid transit and fast intercity trains would be convenient and inexpensive for consumers. Overall, such systems move people with greater energy efficiency, lower maintenance costs, less land use, and greater safety. During the Bush administration vice-president Dan Quayle promoted efficiency in American production to keep us competitive with trading partners. Little attention was paid to getting workers efficiently to production sites.

Public transportation helps poor people the most because they can least afford reliable cars, and they often live where auto fumes are worse than average. But everyone needs less money when second or third cars become unnecessary. This helps people cope with diminished job opportunities in a smaller economy.

As the need for cars lessens, eliminating teenage driving becomes feasible. Many young lives would be saved because automobile accidents are the leading cause of death among people in their late teens. Eliminating teenage driving also makes economic sense for the teens themselves. As the economy contracts there will be fewer jobs for teens at, for example, fast-food restaurants. When government no longer subsidizes meat production, the elevated price of fast food will depress consumer demand.

Rather than work outside the home for money (which is often used at present to support car ownership and use), teenagers can work at home for little or no pay. They can study, cook dinner (the whole family can afford less fast food), maintain the house, and care for younger siblings. Receiving such services free within the family diminishes the whole family's need for money from work outside the home, and orients people more toward family ties.

The quality of life is not diminished as the economy contracts. Transportation is cheaper and safer. Family life is richer, with more family meals. Children are raised by care-givers, parents and older siblings, with whom they have lasting ties, not by underpaid professionals who understandably change employment frequently, often leaving children confused and insecure. As children eat better food and are more psychologically secure, behavior problems at school diminish and the learning environment improves for all.

Energy, Equity, and Population Control

Government policy should generally favor product pricing that reflects full social costs and benefits. This means reducing government subsidies in some cases, such as for transportation by automobile and grain production for export, and increasing subsidies in others, such as for public transportation and sustainable agriculture. In the energy field it means phasing out all direct and indirect subsidies for nuclear power. The price of energy produced by fission should reflect the full costs of health difficulties near mining sites, insurance against reactor mishaps, and the safe burial of nuclear waste. Combined with laws against selling abroad what is too dangerous to use domestically, the nuclear industry will quickly disappear.

Subsidies are in order for home energy improvements, rather than for exploration and use of fossil fuels. Fossil fuel use aggravates global warming, causes health difficulties, and requires defense expenditures to secure supplies. These are all harms and/or costs that home energy improvements reduce as they lessen the need for fossil fuels. During the Carter administration a 15 percent tax credit was given for such improvements. This was discontinued under President Reagan. Yet a Harvard Business School study concluded that for the public to pay the full benefit it receives from home energy improvements, the tax credit should be 60 percent.

When such improvements are in place life becomes less expensive, so people can live with less work outside the home and less income. The economy becomes smaller and people impose themselves less on the nonhuman environment.

Government policy should favor reuse and recycling, for example, with mandatory deposits on bottles and cans, and required use of recycled materials, such as in newsprint. People and businesses should be charged by volume for ordinary trash. The burden on average citizens can be reduced by requiring stores to retain and dispose of packaging that customers choose to leave at the store before taking purchased items home. This will place pressure on producers to reduce packaging.

The full costs of chemicals that harm the environment, such as CFCs that deplete the ozone layer, should be built into products using them, such as air conditioners and packing materials. Some practices that reduce the price of products but pose long-term hazards should be outlawed. For example, nearly one-half of the antibiotics used in the United States are given to farm animals to control disease and lower the cost of meat production. This increases the danger that microorganisms immune to antibiotics will develop and infect human beings. Thirteen thousand people already die annually of bacterial infection in the United States.

Genetic engineering and research should be halted. Although it uses relatively little energy, it promises to alter life fundamentally. Total renunciation deprives some people, those with cystic fibrosis, for example, of potential cures for life-threatening conditions, and we should not view these people as eggs who must be broken in order to make omelettes. But we should equally avoid special treatment for some eggs that jeopardizes many others. If I am correct, success in genetic engineering, to which this cystic fibrosis treatment is tied, will lead to widespread human oppression.

Taxes should be steeply progressive. This provides revenue to fund programs needed to shelter poor people from short-term effects of diminished commerce. Progressive taxation also reduces role models of conspicuous consumption, as there will be fewer rich people. This helps build social expectations around use values. High taxes are said to discourage people who have enough money from working for more. This, too, is good, as rich people will also live more for use values.

As society is more oriented to use values and people need less money to have dignified, comfortable lives, the minimum wage should be increased so that with forty hours work any individual can support at least herself and one dependent without poverty (as society then understands poverty). Mainstream economists claim that increases in the minimum wage discourage employment of people with few skills. I find this another positive effect. Fewer people *should* work extended hours outside the home. Instead, they should play baseball or guitar, or read or paint or gossip. They should develop personal skills and strengthen interhuman ties.

Current technology allows people to produce enough food, shelter, clothing, and material for intellectual, athletic, artistic, and other pursuits that emphasize individual accomplishment and interpersonal relationships. It is simply unnecessary for people to work long hours and increase their domination of nature. Life would contain fewer "toys." But most people, when reflecting on their own priorities, find most meaning in family relationships, friendships, and personal accomplishments (that do not require many "toys"). Winston Churchill said that he knew no one who regretted at the end of his life not having spent more time at the office.

Living more simply with less reliance on exchange values would help our society influence people around the world to curtail population increases. Our current consumption patterns make many people around the world insecure by depriving them of traditional rights to land. People have children in part to regain some sense of security. Reducing our consumption and international trade would encourage return to traditional land tenure systems, as there would be less pressure in many parts of the Third World to use land to gain hard currencies through commodity exports. Personal security would increase and birthrates probably decrease.

Second, when people work to meet local needs with local materials, any disproportion between population size and resource base is apparent. Living under such conditions, many societies controlled their populations for millenia. Modern advances in birth control and epidemiology make this possible without infanticide, war, or plague.

Third, if we lived more simply, countries with high birth rates could no longer claim with justice that our increased consumption, not their increased population, most threatens life on earth. Fourth, many in the Third World believe our attempts to curtail their population increases reflect our greedy desire to reserve more of the earth for ourselves. They realize that they can never have the material riches we enjoy. So why should they give up the one thing they can produce for themselves and enjoy—children—while we give up nothing? The only way to undercut this reasoning is to reduce our own level of consumption.

Finally, we can contribute to declines in worldwide natality by encouraging people in other societies to provide equal education and respect to girls as to boys. In Tamil Nadu, India, noon meals were given to all school children starting in 1975. Most girls attended school at least until lunch was over. When they reached reproductive age around 1990 the birthrate dropped from 35/1,000 to 20/1,000. Due partly to evidence like this, there was general agreement at the worldwide population conference in Cairo in 1994 that empowerment of women is key to controlling population growth.

Concentration on use values furthers women's empowerment. Where exchange values are emphasized, men are usually considered more productive than women because they earn most money. Women do most of the work that receives no pay, such as childrearing, cooking, and cleaning. The increased prestige and purchasing power that exchange relationships give to men increases men's power over women. Accordingly, other things being equal, as people's lives are increasingly oriented toward use values, women's power increases. This can facilitate declining birth rates.

Living with Nature

Personal aspirations and lifestyles will be different when people attempt no longer to dominate nature and one another. Some lifestyle changes are possible already. People can reduce their dependence on exchange values by using public transportation when it is available to avoid having a second or third car. Even with current subsidies, automobile use is expensive. Many people can grow much of their own food, and even those who cannot can save money and improve their health by becoming vegetarians.

People can give up memberships at health clubs and save on lawnmowing charges by cutting their own lawns with old-fashioned reel-type lawnmowers (that have no motors). If more excersize is needed, people can bicycle sometimes and climb (real) stairs.

Spiritual values can be cultivated by identifying with nonhuman aspects of nature and encouraging others, especially young people in one's care, to do likewise. When my granddaughter was born last fall I noted a seedling beneath the maple tree in front of my house. I transplanted it to where it can flourish. My granddaughter will grow up, I hope, recognizing it as her tree.

Unfortunately, what we can do as individuals and families is limited by social context. Effort is needed to change that context through government policies, such as those discussed above.

A few generalizations can be made about life in a society whose economy is oriented to use values. Most people will be less worried and more contented, as there will be less incentive for inventors and advertisers deliberately to sow seeds of worry and discontent to stimulate trade. For example, the more people rely on public transportation, the less they can be pressured by advertisements to spend extra thousands of dollars they can ill afford for a car with leather seats and "performance." Anxiety about paying bills should decrease.

There will be less need to worry about personal security as well. Personal security is threatened today by relative have-nots who find their position intolerable precisely because they lack what society says all must have for a tolerable life. Many turn to drugs to compensate for feelings of alienation and defeat. Lacking money to feed the drug habit, they turn to crime, including violent crime that threatens us all.

In a use-value–oriented economy, people will have what they need and know it. There will be differences in wealth, but these differences will not be tied closely to genuine life chances and self-respect. For example, because cheap, convenient public transportation reduces the pressure on people's lives, fewer will take up drugs and crime.

There need be no lack of excitement in such a society, however. Consider spectator sports. Today professional sports are big business, and I am an avid fan. In a smaller economy people will pay less money to watch sports, because people will generally have and use less money. Professional athletes will not get large salaries, and most will play other roles in the community, as today most poets do more than write and read poetry. Spectator sports will probably then feature more local teams whose players are generally less accomplished than today's big league professionals. But because the teams will be more local, and the players part of the local community in other capacities, the sense of identity among players, teams, and fans is likely to be greater than at present. And with more local teams, more people will be able to watch in person. Anyone who can remember a good high school basketball game will know what I mean. The fun and excitement will be there.

The most important aspect of an economy oriented to use values is that people will have greater opportunity for self-realization. Today, most people's time is dominated by whatever they have to do to make the most money. Many people work two jobs to be able to afford cars with power windows, power seats, sun roof, compact disc player, and so forth. The car must be "loaded," not for transportation, but for self-respect. People have been taught that these things are so valuable that self-respecting folks work hard to acquire them. Others respect people with these things, and consider those without them to be "losers."

In a society oriented toward a smaller economy, people will get the validation we all need from others, and attain self-respect, not through unwelcome work to obtain unneeded "stuff," but directly through joint activities with other people. Because less work for money is needed, people will have more time for less expensive, more family- and friendship-oriented activities, such as bowling, bicycling, hiking, playing cards and other games, participating in amateur theater, playing the piano, and volunteering at nursing homes or at after school programs for children. Because people will not be paid for these pursuits, they will have greater choices about what to do with their time than most people have today. The likely result is that more people will attain self-realization.

The Flight Home

Smoke in the Cabin

When the World Uranium Hearing was over, Christa and Peter Je-cel took Patricia and me for a sightseeing tour of neighboring Austrian towns. Then it was on to Munich and our flight home.

As I sat in the plane, fearful as usual when flying, I began putting together the themes of this book. While making notes furiously, I suddenly noticed an unwelcome smell—smoke. Before I could alert the flight attendant I realized it was just cigarette smoke, and Patricia was coughing because her asthma makes her sensitive.

Because we were many rows away from any smoking section it seemed likely that someone was smoking illegally. Wanting it stopped, I pressed my little call-light button. Five minutes passed without a response so finally I tapped an attendant on the arm as she went by. (This is a true story, reconstructed as accurately as I know how from notes taken at the time.)

After hearing my complaint, she told me there was nothing she could or would do about it. I suggested, in the spirit of good-faith problem-solving, that she make an announcement reiterating the rule that there is no smoking except in designated rows. She said she did not want to raise the issue at all because they already had problems with smoking on the flight. The term *non sequitur* flashed through my mind.

She told me that if I smelled the smoke again, I should push the call-button and someone would come. I pointed out that I had done exactly that some minutes ago, no one came, and the light was on as we spoke. Anyway, what good would it do to draw attention to the problem next time if she would do nothing about it this time?

She suggested that I try to identify the person who is smoking. I pointed out how impractical that was since the person was very likely behind me. If I were to stand in the aisle as smoke patrol I would get in the way of flight attendants serving drinks and doing other work. I suggested again that they make an announcement. Failing that, they could find the smoker themselves, or at least refer the matter to a more senior and responsible flight attendant.

Up the chain of command we went. Several minutes later the head flight attendant came by to explain that they could not make general announcements except during food service because they did not want to awaken sleeping passengers. I pointed out that since my conversation with the other flight attendant, the pilot had made a general announcement of our position. She told me he had done the wrong thing and it would not happen again. This was the first time a flight attendant had ever let me know that she had just reprimanded the captain.

The head flight attendant went on to say that the smoke may very well be coming from rows far away where smoking is permitted because the air filtration system on the relatively new Boeing 767 is not very good. I see, pilots are not permitted to make announcements when they see fit, flight attendants reprimand pilots and tell passengers about it, and new planes of new design are inferior to old planes of older design.

At this point three people sitting nearby told the head flight attendant that they, too, had smelled the smoke. Trying to be reasonable even though I was obviously getting the runaround, I suggested that the next time the pilot was authorized to make an announcement he include a reiteration of the smoking rules.

There never was any announcement, but neither was there more than the slightest hint of smoke during the remaining four hours of the flight. With someone complaining and others willing to confirm the basis for complaint, they decided to do something about a problem whose existence they refused to admit.

I have to thank those flight attendants, because they helped me to crystallize a major thesis of this book. The products of commercial-industrial technology put people in jeopardy. Planes are perfect examples of that. When people are in jeopardy a premium is placed on controlling as many variables as possible, including people. One way of controlling people is to employ hierarchical (bureaucratic) command structures and allow officials to lie whenever they think it is necessary to maintain order.

Choosing What to Believe

I know that most of what I believe comes to me from authoritative sources. I have not discovered, observed, or figured them out for myself. This includes things that I feel I know with certainty, such as the shape of the earth, the cause of tides, the name of the first president of the United States, and my own birthday. (I was there, but could not get a good look at the calendar.) I now discover that lying and covering up are common in the society where I grew up and continue to live. How do I know what to believe?

A major test is correspondence between what I actually observe and what I should expect to observe if the beliefs I have been given are accurate. For example, the story about molecules, air pressure, and air travel is believable because airplanes fly in ways that one would expect if the story were true. Similarly, the tides are what I would expect if they are indeed influenced by the moon. Furthermore, the explanations of these phenomena employ concepts, such as molecules and gravity, featured in many other explanations that seem to work.

There are no guarantees here because it is always possible that what I observe is created by processes different from the ones in the explanations I have been given. But when observations meet expectations, I consider it reasonable to believe in invisible things and forces, such as molecules and gravity.

Disbelief is warranted, I think, when 1) what I observe does not meet expectations created by an explanation, 2) when the explanation does not fit in with other beliefs that seem well founded, or 3) when those supplying the explanation have motives to deceive. These are the elements that make me disbelieve what I was told about the smoke on the plane. The attempt to keep the flight calm gave officials a motive to deceive. The beliefs that new planes are worse than old ones and that airline captains are told by flight attendants when to use the public address system conflict with other beliefs (about technological progress and bureaucratic hierarchies) that seem well founded. The advice that the problem would be taken care of next time if I pressed the call-button did not cohere with my experience, which was that no one came when I pressed the button, and that when I did speak to a flight attendant she was unwilling to do anything.

What does this imply concerning messages in this book, for example, that a growing economy erodes families and promotes frustration, that subduing nature leads to oppressing people, and that helping people requires valuing nature for itself? These messages fly in the face of what we are told (and told and told) is true. What should we believe?

What we are told (and told and told) to believe supports the status quo. This means that many people have motives to deceive (themselves as well as the rest of us). People who are insecure, which is common in today's world, do not want to rock the boat so they have reason to avoid unsettling truths. People with greatest access to the media are generally well served by the status quo, giving them additional motives to deceive themselves and us.

Disbelief is warranted also because many claims make no sense due to conflicts with firmly held convictions. For example, it makes no sense to combat global warming through widespread use of plutonium-producing breeder reactors that create enormous security risks. Similarly, the goal of increasing GDP appears ludicrous when such an increase can stem from people buying medical services to treat illnesses stemming from air and water pollution. We believe firmly that illness and insecurity should be avoided when possible.

The most significant measure is whether or not our observations accord with what received opinion would have us expect. Received opinion says that progress occurs and makes human life better. But violence, homelessness, starvation, insecurity, and the percentage of people institutionalized seem to increase, while close personal ties, personal responsibility, and individual fulfillment all seem to decrease as time goes by. Technology was supposed to provide leisure through labor-saving devices, but most working people seem more time-pressured than ever, whereas people in many traditional indigenous societies have more leisure and less scarcity than us. Observation belies expectations created by received opinion.

On all three measures, then, we have reason to reject received opinion. This book investigates disquieting evidence that protrudes above our culture's reassuring cover-up the way deep rock formations sometimes protrude above the soil.

Denial

It is not easy for me to accept the lessons taught by this evidence and I would not expect it to be easy for others. We cling to early hopes and aspirations on which we base careers and self-concepts. Given my current values and habits, I cannot look forward to living in a society thoroughly transformed to minimize human power over nature. I love travel, especially to Europe, aided by nature-dominating technologies.

But there is no prospect of my living in a thoroughly transformed society. I am called upon merely to promote steps, such as those discussed in the preceding chapter, that effect the needed transformation. These steps should include the progressively altered socialization of succeeding generations so that future people will eventually be fulfilled, like many indigenous people, by local concerns and human relationships. My part in this includes, for example, using more public transportation, recycling, reusing, buying food grown in a sustainable manner, and participating in, while donating funds to, local peace and environmental action groups. Although these activities and many others are possible for me, I cannot say they are all welcome. Even for me the message in this book contains bad news.

Denial is commonly the first stage in the assimilation of bad news, and may be aided by selective perception. Our co-host, Peter Jecel, a man of peace and conscience, was clearly troubled by the Holocaust, which ended when he was only two years old. More than once he raised the topic to explain that it was the work of Germans, not Austrians. Most Austrians opposed unification with Germany, which is why Germany invaded Austria just four days before a scheduled Austrian vote on the issue.

Subsequent reading convinces me that Dr. Jecel is half right. Austrians would have voted against unification with Germany and would never have conducted a genuine Holocaust on their own. But once they were part of Germany, they were more efficiently anti-Semitic, making it soon easier to be a Jew in Berlin than in Vienna. Significant leaders of the Holocaust had Austrian heritage as well, including Eichmann and Hitler.

Dr. Jecel's reaction to the Holocaust is a common form of denial. It readily persists among intelligent, generally informed individuals because denial is accomplished through selective attention. American Jews do this when they stress the democratic nature of the Israeli government and ignore the many forms of discrimination against Arab Israelis. Americans in general do this when they condemn anti-Semitism in Austria in the 1930s, without noticing that Austrian Jews were freer to attend Austrian universities than American Jews were to attend American universities at that time.

Some denials are obvious but symbolically perfect. Manuel Pino spoke at the World Uranium Hearing about the Jackpile Mine near the village of Paguate, New Mexico, where members of the Laguna

Nation live. He noted that blasting at the mine damaged some traditional sandstone pueblo houses, which the mining company repaired by putting stucco in the cracks, as if that repaired structural damage to the house. Another "thing the company did was put paneling on the interior walls of the houses, so that the cracks wouldn't show—like we didn't know there were cracks behind the panel!"

This factual account contains symbolism worthy of fine literature. *Earthen homes* constructed and inhabited by *indigenous people* were damaged by *our culture's explosives* used to *unearth* material that promised to supply *unlimited power over the earth.* The offending corporation applied *palliative cures* designed to *mask the problem* and *deny structural damage.*

Many crime bills are analogously predicated on denial. They promise to get tough on criminals by increasing penalties, expanding use of capital punishment, reducing judicial discretion in sentencing, and requiring more minors to be tried as adults. Use of these approaches has been expanding for decades now, while violence, crime, and the percentage of the population incarcerated increase regularly. It is time to recognize that problems of violence and crime are structural. Our culture's house needs fundamental repair.

Denial of this book's message is tied to denial of death. For many people the dream of controlling nature in the human interest includes faith in medical progress, which joins or replaces the New Testament's promise of eternal life. The American Heart Association thinks that with enough effort, death by heart disease can be reduced. The American Cancer Society uses hope of curing cancer in our lifetime to solicit research funds. Either these medical groups have some preferred cause of death in store that they have not yet mentioned, or they are tempting us to seek earthly immortality. The latter is classic denial.

In the last forty years age-standardized cancer rates have gone up 43.5 percent in the United States. Current increases in lung and breast cancer are epidemic. Exposure to radiation, agricultural chemicals, and industrial wastes are implicated. Except for several types of cancer, survival rates among cancer patients have not improved for decades. Media coverage of cancer cures is like wood paneling covering structural damage to earthen homes—It assists denial.

Genetic research to "save the lives," for example, of people with cystic fibrosis is the same. The rationale of "saving lives" is excellent evidence that this project of controlling nature is predicated on unrealistic denial of death.

Overcoming this denial is probably the deepest change in thinking needed by many of us who would revere nature as it is and think of nonhuman environmental constituents as partners in social exchange. Only when we accept timely death will we attain life as plain members and citizens of the biosphere.

Sources

Opening Quotes

Joseph Goebbels is quoted in Zigmunt Bauman, *Modernity and the Holocaust* (Ithaca, N.Y.: Cornell University Press, 1989), p. 71. Francis Bacon is quoted in Carolyn Merchant, *The Death of Nature* (New York: Harper and Row, 1980), p. 172. Alvin Weinberg is quoted in *Taking Sides: Clashing Views on Controversial Environmental Issues*, Theodore D. Goldfarb, ed. (Guilford, Conn.: Dushkin Publishing, 1991), p. 144.

For reproductive cancer among Navaho teens, see Dick Russell, "Environmental Racism," *The Amicus Journal* (Spring 1989), p. 24.

George Blondin, Lorraine Rekmans, and all others who testified at the World Uranium Hearing are quoted in *Poison Fire/Sacred Earth*, Sibylle Nahr and Uwe Peters, eds. (Munich: The World Uranium Hearing, 1993). Page numbers are located through the speaker's name in the table of contents. Blondin's quote is from p. 83, Rekmans' are from p. 66, and Yazzie's is from p. 26.

Introduction

All biblical references are taken from *The Scofield Reference Bible*, Rev. C. E. Scofield, D.D., ed. (New York: Oxford University Press, 1945). This quote is Gen. 1:28.

For Aristotle and Cicero, see Ian G. Barbour, *Technology, Environment, and Human Values* (New York: Praeger Publishers, 1980), pp. 14–15.

For readable accounts of Kant, Hobbes, and most other philosophers cited in this book, see Peter S. Wenz, *Environmental Justice* (Albany, N.Y.: State University of New York Press, 1988).

John Passmore, *Man's Responsibility for Nature* (New York: Scribner's, 1974).

William F. Baxter, *People or Penguins: The Case for Optimal Pollution* (New York: Columbia University Press, 1974), quoted in Bill McKibben, *The End of Nature* (New York: Anchor Books, 1990), pp. 151–152.

Jamake Highwater, *The Primal Mind: Vision and Reality in Indian America* (New York: Meridian, 1981), pp. 77–78.

The classic source for the land ethic is Aldo Leopold, *A Sand County Almanac* (New York: Oxford University Press, 1949).

Chapter 1

Father Thomas Berry's views are found in *The Dream of the Earth* (San Francisco: Sierra Club Books, 1990), pp. 81 and 126.

Barbara Tuchman is quoted from *A Distant Mirror* (New York: Ballantine Books, 1978), p. 104 (on the plague as divine chastisement), p. 211 (on women), pp. 338–339 (on Wyclif's heresy), p. 590 (on the inquisition), and p. 202 (on infidels).

The fourteenth-century preacher was quoted in Carol Tavris and Carole Wade, *The Longest War* (Orlando, Fla.: Harcourt Brace Jovanovich, 1984), p. 8.

Zigmunt Bauman's *Modernity and the Holocaust* is referenced above. The quote is from p. 37.

Federic Duncalf, "The Councils of Piacenza and Clermont," in *A History of the Crusades: The First Hundred Years,* Marshall W. Baldwin, ed. (Madison, Wis.: University of Wisconsin Press, 1969), pp. 243–244.

Juan Gines de Sepulveda is quoted in *1492: Discovery, Invasion, Encounter,* Marvin Lunenfeld, ed. (Lexington, Mass.: D.C. Heath and Company, 1991), p. 220.

Descartes's views on nonhuman animals are found in *Animal Rights and Human Obligations,* Tom Regan and Peter Singer, eds. (Englewood Cliffs, N.J.: Prentice Hall, 1976), pp. 64–65. For a critique of Descartes's views, see Tom Regan, *The Case for Animal Rights* (Berkeley, Calif.: University of California Press, 1983). Information on the treatment of nonhuman animals can be found in Peter Singer, *Animal Liberation* (New York: Avon Books, 1975, 1993).

Chapter 2

Clement A. Tisdell, *Natural Resources, Growth, and Development: Economics, Ecology, and Resource-Scarcity* (New York: Praeger Publishers, 1990), pp. 1–2.

The British Columbia Indian agent is quoted in Peter J. Usher, Frank J. Tough, and Robert M. Galois, "Reclaiming the Land: Aboriginal Title, Treaty Rights and Land Claims in Canada," *Applied Geography* 12 (1992), p. 121.

See "Whose Common Future?" *The Ecologist* 22:4 (July/August 1992), pp. 132–133 (on enclosures) and p. 170 (on increasing income gaps between rich countries and poor ones).

For critical reviews of current free trade agreements, see Kristin Dawkins and William Carroll Muffett, "The Free-Trade Sellout," *The Progressive* 57:1 (January 1993), pp. 18–20; Nicholas Hildyard, "Maastricht: The Protectionism of Free Trade," *The Ecologist* 23:2 (March/April 1993), pp. 45–51; and Tim Lang and Colin Hines, *The New Protectionism* (London: Earthscan Publications, 1993). See *The New Protectionism* also for the failure of trickle-down economics in the United Kingdom (pp. 78–79), and for increases in the Third World debt (p. 13).

For information on European colonization of the Americas, see Paul Johnson, *The Birth of the Modern* (New York: HarperCollins, 1991), especially pp. 209, 220, and 249.

For information on slaughter in the colonies, see Ward Churchill, "Deconstructing the Columbus Myth," in *Confronting Columbus,* John Wewell, Chris Dodge, and Jan DeSirey, eds. (Jefferson, N.C.: McFarland Publishers, 1992), pp. 149–158.

The history of injustice in Central America is drawn largely from Walter LaFeber, *Inevitable Revolutions* (New York: W. W. Norton, 1984), especially pp. 54, 70–71, 81, and 115. See the same work pp. 111–126 and 204 for U.S. help in maintaining dictatorships in Latin America.

For quotes and figures on the increasing gap between the rich and poor internationally, see Paul Kennedy, "Preparing for the 21st Century: Winners and Losers," in *Global Issues 94/95,* Robert M. Jackson, ed. (Guilford, Conn.: Dushkin Publishing, 1994), p. 18.

On child prostitution, see Aaron Sachs, "The Last Commodity: Child Prostitution in the Developing World," *World Watch* (July/August 1994), pp. 24–30, especially p. 26.

On issues of hunger see Frances Moore Lappe and Joseph Collins, *World*

Hunger: Twelve Myths (New York: Grove Weidenfeld, 1986), especially pp. 85–89. For helpful statistics, see *Poverty and Hunger: Issues and Options for Food Security in Developing Countries* (Washington, D.C.: World Bank, 1986). This report gives the figure of 40,000 children dying unnecessarily every day.

For issues of gender inequality, see Jodi L. Jacobson, "Closing the Gender Gap in Development," in *State of the World 1993,* Lester R. Brown, ed. (New York: W. W. Norton, 1993), especially pp. 62–67.

Chapter 3

For standardization at McDonalds, see Barbara Garson, *The Electronic Sweatshop* (New York: Penguin Books, 1988), chapter 1.

For information on nineteenth-century slavery in the United States, see Paul Johnson, cited above, pp. 309–313.

For the impossibility of everyone having U.S. material standards of living, see Donella H. Meadows, Dennis L. Meadows, and Jorgen Randers, *Beyond the Limits,* (Post Mills, Vt.: Chelsea Green Publishing, 1992), especially pp. xiii–xiv and 38.

The editorial is in *The Wall Street Journal Europe* (February 15, 1993).

Information on GATT is in Kristin Dawkins and William Carroll Muffett, as well as in Tim Lang and Colin Hines, both cited in sources for chapter 2. In Lang and Hines, see especially pp. 48–49 and 52.

For sociobiology against women, see Daniel G. Freedman, *Human Sociobiology* (New York: The Free Press, 1979), especially pp. 12 and 19, and for infants' preference for female voices, p. 145. A good critique is in Barry Schwartz, *The Battle for Human Nature* (New York: W. W. Norton, 1986), especially p. 101 (on asexual reproduction), pp. 193–195 (on the role of constraints in evolutionary development), and p. 211 (on societies that practice infanticide but adopt the children of others).

Chapter 4

The quote from Ernest Gellner is found in *Plough, Sword, and Book* (Chicago: University of Chicago Press, 1990), pp. 186–187.

Information on the Holocaust is drawn from William S. Shirer, *The Rise and Fall of the Third Reich* (New York: Simon and Schuster, 1959), especially pp. 1275–1285; Zigmunt Bauman, *Modernity and the Holocaust* (Ithaca,

N.Y.: Cornell University Press, 1989), especially pp. 71 (for Goebbels quote), p. 187 (for Himmler quote), and p. 14 (for Weber quote); and Bruce F. Pauley, *From Prejudice to Persecution* (Chapel Hill, N.C.: University of North Carolina Press, 1992), especially pp. 22 and 289.

Information on Beirut is from Thomas L. Friedman, *From Beirut to Jerusalem* (New York: Anchor Books, 1990), pp. 42–43.

Aldo Leopold's optimistic view of moral progress is in *A Sand County Almanac,* cited above, p. 237. J. Baird Callicott's views appear in "The Conceptual Foundations of the Land Ethic," in *Companion to a Sand County Almanac,* J. Baird Callicott, ed. (Madison, Wis.: University of Wisconsin Press, 1987), p. 188.

For information on racial disparities, see Harold Henderson, "A House Divided," *Illinois Times* (May 27–June 2, 1993), pp. 8, 10, and 11; and Thomas Atkins, "Segregated Springfield," same issue, p. 9.

National Public Radio covered Human Rights Watch Report in December, 1993. Tom Sjelten, *Sarajevo Daily* (New York: HarperCollins, 1995) makes the same point.

On the worsening condition of children worldwide, see Germaine W. Shames, "The World's Throw-Away Children," in *Global Issues 94/95,* pp. 229–32. Quotes are from pp. 229 and 230. The book is cited in sources for chapter 2.

Information on the Carnegie Foundation Report concerning the worsening conditions for children in the United States was drawn from a report on National Public Radio, *Morning Edition,* April 13, 1994.

Chapter 5

See Nicholas Lenssen, "Confronting Nuclear Waste," in *State of the World 1992,* Lester R. Brown, ed. (New York, W. W. Norton, 1992), p. 48 (for recent recognition of the toxicity of radiation), p. 49 (for predicted deaths from Chernobyl disaster), pp. 46–47 and 60 (for backlog of high-level nuclear waste), p. 67 (for Yucca Mountain being claimed by the Western Shoshone Indians), pp. 55 and 58 (for geological uncertainty), p. 61 (for copper drum disposal method), p. 58 (for early scientific warnings concerning nuclear wastes), and p. 62 (for Japan disposing of nuclear wastes in China).

For Martin Gardner's research, see Bill Keepin, "Children of Plutonium," *Nuclear Guardianship Forum* (Issue 2, Spring 1993), p. 3.

See Miles Goldstick, "The Ability of Alpha Radiation to Penetrate Human Skin," *WISE* (World Information Service on Energy) (September 1992), p. 6, on the health of Czecholsovakian miners.

For illnesses among French uranium miners, see *Info-Uranium #48* (1991), p. 4.

For India's nuclear program being the most independent in the Third World, see Christopher Flavin, *Nuclear Power: The Market Test,* Worldwatch Paper 57 (Washington, D.C.: Worldwatch Institute, 1983), p. 53.

See Ward Churchill and Winona LaDuke, "Native America: The Political Economy of Radioactive Colonialism," in *Critical Issues in Native America,* vol. II, Ward Churchill, ed. (Copenhagen: International Group for Indigenous Affairs, 1991), p. 46 (for 93% of Anaconda workforce being from the pueblo), pp. 32–34 (for most other information in this chapter on mining conditions, including conditions at the Shiprock mine), and p. 41 (for the Nixon administration plan for a national sacrifice area).

Martine Deguillaume testified at the Hearing that the French government did not test milk samples for 40 years.

See also Alan Thein Durning, "Nowhere to Run: Native Americans Stand Their Ground," *World Watch* (November/December 1991), pp. 10–17, for more on native people's attachments to their land.

For biological activity of fission products, see *A Nuclear Power Primer* (League of Women's Voters Education Fund, 1983), p. 27.

For safety problems in the French nuclear program, see Christopher Flavin, "The Case Against Reviving Nuclear Power," *The World Watch Reader,* Lester R. Brown, ed. (New York: W. W. Norton, 1991), p. 210. Also on this matter, see Marnie Stetson, "France's Tottering Nuclear Giant," *World Watch* (January/February 1991), p. 38.

For percentage of radioactivity coming from high-level waste, see William Poole, "Gambling with Tomorrow," *Sierra* (September/October 1992), p. 53. This is generally a good source for issues surrounding long-term disposal of nuclear waste.

Chapter 6

See Christopher Flavin, "The Case Against . . . ," cited above, p. 213 (for variety of reasons given over the years to justify nuclear power), p. 211 (for the cost of the Chernobyl clean-up), pp. 205–206 (on the Shoreham Nuclear Power Plant), and p. 217 (on the global warming rationale).

See *A Nuclear Power Primer,* cited above, p. 13 (on the Price-Anderson Act) and pp. 67–69 (on the Non-Proliferation Treaty).

For government subsidies to nuclear power, the cost of nuclear energy, and WPPSS, see Christopher Flavin, *Nuclear Power: The Market Test,* cited above, pp. 18, and 40–41.

For financial difficulties of French nuclear program, see Marnie Stetson, cited above, p. 38.

See John Tagliabue, "A Legacy of Ashes: The Uranium Mines of Eastern Germany," *The New York Times* (March 19, 1991), on the cost of cleaning up East German uranium mines.

See Nicholas Lenssen, cited above, especially p. 52, on Wismut mine cleanup costs, costs of burying all current high-level nuclear waste, and problems of decommissioning power plants.

See David Gow, Paul Brown, and Simon Beavis, "Double Blow to Thorp Plant," and Paul Brown and Simon Beavis, "The Key to Thorp's Nightmare Scenario," *The Guardian* (October 20, 1993), pp. 1 and 2, for the expense and increase of waste by volume of reprocessing nuclear waste.

For the scarcity of uranium, a critique of the global warming rationale, and the nature of FBRs, see Nigel Mortimer, "Nuclear Power and Carbon Dioxide," *The Ecologist,* 21:3 (May/June 1991), pp. 130–131.

For the toxicity of plutonium, the danger of shipping it, the problem of diversions, and the dangers of FBRs, see Allen V. Kneese, "The Faustian Bargain," in *Energy and the Environment,* Harold Wolozin, ed. (Morristown, N.J.: General Learning Press, 1974), pp. 121–122.

The poll of American attitudes about North Korea was a USA Today/CNN Poll, reported on CNN in November 1993.

For how Jews were blamed for their landlessness, see Bruce F. Pauley, *From Prejudice to Persecution* (Chapel Hill, N.C.: University of North Carolina Press, 1992), p. 216.

Chapter 7

See Richard B. Lee, *The Dobe !Kung* (the new edition is *The Dobe Ju/'hoansi*) (New York: Holt, Rinehart, Winston, 1984, 1993), p. 50 (on hoarding meat) and p. 53 (on not working many hours per week).

See Richard B. Lee, *The !Kung San* (New York: Cambridge University Press, 1979), p. 372 (on the number of homicides among the Ju/'hoansi), p. 399 (for the quote about the externalization of violence), p. 326 (on birth spacing), p. 322 (on population increase with sedentary life), and chapter 10 (on food abundance).

See David Maybury-Lewis, *Millennium: Tribal Wisdom and the Modern World* (New York: Viking Penguin, 1992), p. 85 (for quote from the Gabra) and pp. 65–68 (on indigenous exchange relationships).

For violence of the Semai fighting with British forces, see Robert Knox Denton, *The Semai* (New York: Holt, Rinehart and Winston, 1968), p. 58.

For issues of population control and population increases, see "Whose Common Future?" *The Ecologist* 22:4 (July/August 1992), pp. 39–41 and 170–171.

See Marshall Sahlins, *Stone Age Economics* (Hawthorne, N.Y.: Aldine Publishing Co., 1972), p. 203, on the ability of pastoralists to control their populations.

See Orlando Behling and Chester Schriesheim, *Organizational Behavior* (Boston: Allyn and Bacon, 1976), pp. 59–60, on Maslow's hierarchy of needs.

Food pantry needs in 1994 were reported on CBS Evening News, December 19, 1994.

Chapter 8

See Thomas W. Overholt and J. Baird Callicott, *Clothed-in-Fur and Other Tales: An Introduction to an Ojibwa World View* (New York: University Press of America, 1982), p. 1 (on the nature of a world view) and pp. 146 and 151–152 (for Objiwa views on reciprocal exchange with animals).

See Nurit Bird-David, "Beyond 'The Original Affluent Society'," *Current Anthropology* 33:1 (February 1992), pp. 31–32 (for indigenous assumption of abundance), pp. 25 and 29–30 (for social view of the environment and its constituents among foragers), and p. 29 (for quote about being children of the forest).

See Marshall Sahlins, cited above, p. 36 (for indigenous inability to imagine anyone starving to death), and pp. 37 and 87–88 (for the lack of indigenous poverty).

See Mary Midgley, *Evolution as a Religion* (New York: Methuen, 1985), pp. 14 and 59, on the nature of meaning in life.

See David Maybury-Lewis, cited above, pp. 137–142 (for the importance of ceremony and sense of security and role consistency in indigenous life), p. 123 (for quote from Paul Radin), pp. 53–54 (on Xerente Indians' relationship to the peccary), p. 55 (on Makuna views about fish), pp. 68–73 (on the Weyewa non-commercial exchange), and p. 81 (for Gabra non-commercial exchange).

For the environmentally benign effects of indigenous societies, see Alan Thein Durning, "Supporting Indigenous People," in *State of the World 1993*, Lester R. Brown, ed. (New York: W. W. Norton, 1993), p. 85.

Chapter 9

For the GDP stimulus of oil spills and air pollution, see Sandra Postel, "Toward a New 'Eco'-Nomics," *World Watch* (September/October 1990), p. 22.

I owe the example of running water in Mexican villages, and the distinction between individuality and individualism, to anthropologist colleague Jim Stuart.

I owe the phrase "Invention is the mother of necessity" to my colleague Maria Mootry.

Chapter 10

Vernon W. Ruttan, *Sustainable Agriculture and the Environment: Perspectives on Growth and Constraints* (Boulder, Colo.: Westview, 1991). For subsidies to the meat industry, see "Matters of Scale," *World Watch* (July/August 1994), p. 39.

Lang and Hines, *The New Protectionism*, cited in sources for chapter 2, contains information on Brazilian inequality (p. 136), on Brazilian food production for export while its own people are malnourished (p. 96), and the anti-environmental aspects of international trade agreements (pp. 64, 68–71, and 129). The longer quote about Brazil's agriculture for export is from David Morris, "Free Trade: the Great Destroyer," *The Ecologist*, 20:5 (September/October 1990), p. 191. Information on royalties to be paid for the use of seeds is from John Vidal, *The Guardian* (October 1, 1993).

The Harvard study on the benefits of tax credits for home energy improvements is cited in Ian Barbour et al., *Energy and American Values* (Buffalo, N.Y.: Praeger Publishers: 1982), pp. 110–111.

The use of anti-biotics for farm animals was discussed on National Public Radio, *Earth Matters*, March 25, 1995.

Natality decrease in Tamil Nadu, India is discussed in Robert Goodland, "South Africa: Environmental Sustainability Needs Empowerment of Women," in *Faces of Environmental Racism*, Laura Westra and Peter S. Wenz, eds. (Lanham, Md.: Rowman and Littlefield, 1995).

The Flight Home

For Austria's role in the Holocaust, see generally Bruce F. Pauley, cited above.

For increases in rates of cancer and cancer deaths, see Samuel S. Epstein, "Profiting from Cancer: Vested Interests and the Cancer Epidemic," *The Ecologist* 22:5, pp. 233–240.

Index